The Evolution of Racism

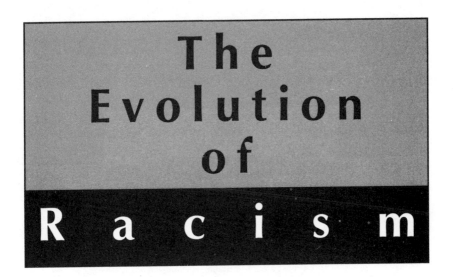

The Evolution of Racism

*Human Differences and
the Use and Abuse of Science*

Pat Shipman

HARVARD UNIVERSITY PRESS
Cambridge, Massachusetts

Copyright © 1994 by Pat Shipman
All rights reserved
Printed in the United States of America

Originally published by Simon & Schuster, Inc., New York, New York

First Harvard University Press paperback edition, 2002

Designed by Deirdre C. Amthor

Library of Congress Cataloging-in-Publication Data

Shipman, Pat.
The evolution of racism: human differences and
the use and abuse of science/Pat Shipman.
p.   cm.
Includes bibliographical references and index.
1. Racism—History. 2. Evolution (Biology) I. Title.
HT1507.S48   1994
305.5′009—dc20   94-4459
                                                  CIP
ISBN: 0-674-00862-6 (pbk.)

Acknowledgments

No BOOK IS ever written without assistance, and this is no exception. Many individuals have helped by granting me interviews, sending me reprints, reading drafts of chapters, or discussing topics with me, thus influencing and clarifying my sometimes muddled thoughts. I have not always agreed with their perspectives nor do they all endorse mine, but I assure them that I have listened sincerely and carefully. These include C. Loring Brace, Peter and Ginger Ross Breggin, Matt Cartmill, Margarete Heck and Bill Earnshaw, Heinz Heck, F. Clark Howell, Roger and Gail Lewin, Paul McHugh, Sidney Mintz, Ashley Montagu, Lita Osmundsen, Christopher Stringer, Mara Connolly and Pete Taft, Pamela Maupin and Bill Tew, Alan Thorne, Erik Trinkaus, Ronald Walters, David Wasserman, and Milford Wolpoff.

Other friends and relations have also helped to keep me on an even keel through the writing of this book: Ginny Armstrong, Sharyn Marzullo, Diane Suiters, Chris Wadman, Simon and Shellene Walker in particular. I would also like to thank my editor, Bob Asahina, his assistant, Sarah Pinckney, and my copy editor, Phil James, for cogent comments, patience, and intelligent support.

Last but not least, my husband, Alan Walker, has as always been a tremendous help on matters both intellectual and emotional; my debt is greater than can be expressed. Finally, I thank my horse, Agincourt, and my cat, Amelia, for lending their singular perspectives and inimitable humor to this endeavor.

FOR ALAN

Contents

"To ignore race and treat an individual as an individual is the spring of justice and the river of hope."

—Leigh Van Valen, 1966, "On Discussing Human Races."

Prologue

W$_{\text{E ARE A}}$ homocentric species, strangely self-absorbed.

Questions of *what* and *who* we are supersede all others in importance. From the very beginning, Darwin's evolutionary theory has been irrevocably intertwined with the contentious matters of the position of humans and even the very definition of humanity. They confound us still.

And Darwin understood the crux of the problem: "It is not my intention here to describe the several so-called races of men; but I am about to inquire what is the value of the differences between them under a classificatory point of view and how they have originated."[1] That is it: the *value of the differences* among them—among us.

This is the story of some of those who have struggled to define that value—who have been puzzled, tormented, enriched, and sometimes elated by their hard-won views. It is the story of a problem that has grown and developed, as evolutionary theory itself has migrated freely from an obscure academic argument to a theory that pervades modern society. In the process, both the theory and our views of ourselves have been filtered through the minds of men and women of all stations and abilities, and have been altered unutterably by the process. Careers have been made and lost, wars have been provoked and fought, lives dedicated or sacrificed, societies healed or rent asunder over the value of the differences.

It begins, as does so much, with Darwin. It ends with ourselves.

Part I
Evolution

One Long Argument: 1857

CHAPTER 1

H E NEVER MEANT to start trouble.

In fact, he hated disagreements, quarrels, emotional disruptions of any sort. They gave him a sick headache or an upset stomach. Only Emma really understood, and soothed him with her quiet ways and her efficiently run household, full of happy children. Each morning after his solitary walk, breakfast, and an hour and a half's work, she would bring his mail to him as he sat in his favorite armchair, with the board across the arms that served as a desk. Then she would settle down near-by to read family letters aloud or quietly attend to needlework. She always listened patiently if he had an interesting letter—perhaps a puzzling response to one of the endless queries with which he pestered friend and stranger alike—and sympathetically if he had complaints about his physical ailments. Indeed, as one of their daughters observed, "Her whole day was planned out to suit him, to be ready for reading aloud to him, to go on his walks with him, and to be constantly ready at hand to alleviate his daily discomforts."[1] She was the perfect help-meet.

Though once he had eagerly sought the company of other gentlemen interested in science, a reticence had come on him gradually which now set the pattern of his daily life. The price of intellectual stimula-tion—of a lively evening at the Geological Society, for example—was always a subsequent day of illness and discomfort. It was better to stand apart, to learn of new ideas by letter or written papers that he could study slowly at his leisure, thinking out their premises and consequences before he ventured an opinion or a comment. All those quick-witted chaps, ready to argue a new theory only moments after they had heard of it—no, that was not his way. He was a bit of a plodder now, slow on the uptake and hesitant in his speech. He had no gift for the clever turn of phrase and the piercing quip that thrust to the heart of an issue, no taste for intellectual combat, like young Huxley. What he had was a ravenous intellect, greedy for facts and understanding. He compiled long lists of specific questions on various subjects and saved them until he could beg half an hour's time from someone who was an authority.

On the whole, he preferred to be left alone at Down House, reading and working in the morning and going for midday strolls along the Sand Walk with one of the children. He liked to point out to their eager, young eyes the different plants, insects and animals and their habits. In the afternoon, he read *The Times*, wrote innumerable letters of inquiry, rested, worked for another hour or so, and settled on the sofa to listen to Emma's gentle voice reading a novel. He often lamented that he got in only a few hours' good work on his theories in a week—though he did not count time spent reading articles, answering or initiating correspondence, or doing experiments as work, which explains his otherwise miraculous productivity. And sometimes his colleagues came to visit—just his friends, one or two at a time, no crowds to overstimulate, only someone to "pump" (as he called it) with his catalogs of questions—but even this was exhausting.

His name was Charles Darwin.

He was about to start a "long argument"[2] that would persist far longer than he could have imagined—for decades, maybe centuries to come. It would shake the moral underpinnings of Western civilization, challenging the traditional, Christian belief in a single episode of creation of a static, perfect, and unchanging world. If the world were not created perfect, then there was no implicit justification for the way things were, the way things had always been—and that thought has lain at the heart of all of the world's great sociopolitical and many of its scientific revolutions. Darwin's long argument would provoke fervor and rancor as few debates ever have; it would demand the utmost in courage and intellectual honesty that each succeeding generation of scientists could muster.

The argument would focus first on the theory of evolution, on a view of the way the world worked and species originated that was singularly Darwin's own. But in an astonishingly short time, the argument would shift from the theory itself to its application to human beings, and thence to the value of the differences among the human races. This would provoke a long and painfully introspective examination of our species and its amazing variability that is not finished yet.

Races, or at least groups of people with different cultures, languages, and appearances, were still being discovered by explorers from the European nations. It was the age of imperialism and most non-Europeans were regarded, even by Darwin, as "barbarians"; he was astonished and taken aback by their wildness and animality.[3] The differences among humans seemed so extreme that the humanity (and modernity) of some living groups was scarcely credible. The diversity of humankind was as much a cause for wonderment and puzzled head-shaking as the presence of snow-capped mountains on the African equator. The full spectrum of human appearance and behavior was like a massive continent only one shore of which was familiar to most Europeans.

To be responsible for such ferment was a situation in which Darwin had hoped never to find himself; he was no social rebel, no proud defiant. Yet, when it arose, he found the courage to speak and write despite the fact that he was appalled by the deep discord that he anticipated his words and ideas would spawn.

His was a difficult road to travel. Darwin had spent a lifetime doing the expected, never stepping outside the bounds of the acceptable. Even his early aimlessness, which had caused his father to despair of his ever being useful, was of a thoroughly predictable type. And a wastrel he might have been, as his father had ominously predicted, except that he discovered geology at Cambridge. Otherwise, Darwin would doubtless have been an entirely amiable sort of ne'er-do-well: a hunting-shooting-and-fishing son of a well-to-do father—a father who would pick a sensible cousin for his son's bride and whose wealth would provide that son with a means to keep his large family. Darwin would never have been one to challenge society purposefully, to overturn mores and flaunt public morality for the sheer excitement of defiance. To him, heterodoxy was an awful sin.

His aversion to facing his theory and its implications was so great that he had difficulty admitting to it, even to his friends. Year after year, he circumvented any bald, written statement of his beliefs, disguising them behind needless words and dithering clauses. He felt compelled to distance himself from his ideas even as he was driven to bring them forth and try them out.

"I have been now ever since my return [in 1836, from the voyage of the *Beagle*] engaged in a very presumptuous work, and I know no one individual who would not say a very foolish one," Darwin finally dared to write to his close friend, the botanist Joseph Hooker, in 1844. As he continued this letter, his increasingly convoluted sentences betrayed his close-coupled need to acknowledge his thoughts, like bastard off-spring, and his anguish at their anticipated reception. He explained to Hooker:

> I was so struck with the distribution of the Galapagos organisms, &c. &c., and with the character of the American fossil mam-mifers, &c. &c., that I determined to collect blindly every sort of fact, which could bear any way on what are species. . . . At last gleams of light have come and I am almost convinced (quite con-trary to the opinion I started with) that species are not (it is like confessing a murder) immutable.[4]

At about the same time, he admitted reluctantly to his friend Asa Gray, another botanist: "As an honest man, I must tell you that I have come to the heterodox conclusion that there are no such things as independent-ly created species—that species are only strongly defined varieties. I know that this will make you despise me,"[5] he concluded miserably.

Species were not, then, fixed, unchanging, constant since their per-fect creation. Was this idea such a sin, such a violence to contempo-rary thinking as to be called a *murder*? His words seem hyperbolic, but events would prove they were not. The vision of the world as once-cre-ated and henceforth the same, the view of God as the watchmaker whose clever machinery continued to simply tick over, were more than comfortable; they were the very basis upon which society operated. Small wonder that Darwin's worst fears of uproar and dissent were to be realized.

Little more than a year before his book was published, he was still hesitantly sketching these ideas to others in this way, hoping for sup-port and assistance without daring to explain the theory fully, unable either to suppress his thoughts or to espouse them without a show of diffidence. But his dismay at that point was nothing compared with his distress at the time of publication of *The Origin of Species* in Novem-ber, 1859. Beforehand, he caught a glimmering of the trouble his ideas would cause—and did not look forward to it happily. He tried as much as possible to avert public outcry.

"As this whole volume is one long argument," Darwin sighed in the last chapter of *The Origin of Species*, knowing others would argue back, "it may be convenient to the reader to have the leading facts and infer-ences briefly recapitulated."[6]

They were these. Given that species evolved over time, Darwin strove to deduce a means by which this descent with modification might occur. He cleverly named his mechanism *natural selection* in an explicit parallel to the humanly induced selection that was familiar to anyone acquainted with the breeding of domestic stock, dogs or horses, for this was the insight that showed Darwin how the whole mechanism of evolution might operate. Although nothing was known of modern genetics, DNA, or chromosomes, it was apparent to any countryman that offspring often inherited the characteristics—physical and mental—of their parents. Darwin's most convincing proof of his theory, to many readers, was the evidence of breeding practices that were put into use every day, with important consequences.

In effect, Darwin said this: As some horses are selected by their owners for breeding because they are faster runners, and some cows for their higher yield of milk, so too do the different variations in physique or ability among a wild species sometimes enhance or damage the reproductive success of particular individuals. It was a commonplace in Darwin's day that any good horseman looks for the best stallion to breed to his prize mare, hoping that their get, or offspring, will enjoy the best qualities of each; advertisements for the services of particular stallions appeared regularly in newspapers in the eighteenth, nineteenth, and early twentieth centuries. In fact, so irrefutable was the benefit of selective breeding (and stud books, in which careful records were kept of crosses) that English mares had been segregated from stallions except for planned breedings since the year 1130.[7]

Natural (as opposed to artificial or man-made) selection might work similarly, according to Darwin: by imparting a reproductive advantage to some of the naturally occurring variations in form or ability. This was more than a metaphor, it was an intuition that domestic breeding practices and natural selection were the same phenomenon differing only in the *identity* of the selector (humans, on the one hand, and the exigencies of survival of the fittest, on the other) and the *scale* of the effects produced by selection. Because of the vast time periods over which natural selection operated, small differences among the offspring would become heightened or exaggerated and the numbers of survivors of those who bore advantageous traits would become greater. Eventually, such traits will predominate in any population of a single species, enabling it to become better and better adapted to its environment and circumstances. The logic of it was excellent. It was just like pigeon breeding, writ large.

But why did Darwin believe that species were mutable? During the voyage of the *Beagle*, between 1831 and 1836, Darwin's peace and tranquility were disturbed by a haunting idea, a growing vision that was born of his first confrontation with the wide, strange world. His main companions were the captain, Robert FitzRoy—a young aristocrat with whom Darwin shared a cramped cabin—and Syms Covington, a teenaged sailor whom Darwin had taken on and trained as his assistant and personal servant. Of course, in another sense Darwin had his books for company. Between them, FitzRoy and Darwin boasted a library of 246 volumes, including (on Darwin's side) John Milton's *Paradise Lost*, to nurture his Christian soul, and Charles Lyell's newly published *Principles of Geology*, to stimulate his geological mind.

Then young and strong, Darwin had walked or ridden on horseback for miles at every stop on the voyage, covering terrain rugged or smooth, climbing every mountain he saw, collecting every organism he or Covington could get in a net. It was an orgy of acquisition and observation. He was, in many senses, a tabula rasa, a blank page awaiting the copperplate hand of life.

"One great source of perplexity to me," he wrote to his former Cambridge professor, botanist J. S. Henslow, "is an utter ignorance whether I note the right facts, and whether they are of sufficient importance to interest others."[8] His characteristic solution to the problem was simply to note *everything*. He was not, as yet, a man with an idea, searching for evidence to support (or disprove) it. He was simply blotting paper, soaking up life's ink.

He saw that, in a single geographic region, many animals, plants, or birds bore such close resemblances to one another that they were surely closely related. And yet, if they lived under even slightly varying conditions, physical differences within the broad similarities could be observed, as with the famous Galapagos finches and tortoises. While the resemblances among the finches or tortoises were striking, populations on adjacent Galapagos islands varied in subtle ways: by having a longer beak or a stouter one, a higher-domed carapace or a flatter shell. Could God's whimsy explain this repeated pattern of variability within overall similarity? Or was something else at work? Suppose each population included natural variability in beak or shell, hoof or feather, that would render some better suited than another to a specific habitat. Since the numbers of offspring of any species far exceed the numbers that can be supported by a particular habitat, some must die: those less suited. Thus, Darwin realized slowly, those best suited to the local circumstances would survive best and leave more offspring.

Darwin could see that his natural selection would fuel the process of adaptation to differing habitats by differential survival and reproduction—this group living in a more arid area than that one, or the one population coping with denser vegetation than the other. Adaptations would, of course, be manifested physically, in longer or shorter limbs, thicker or thinner fur, more pointed or blunter beaks, and so on, each set of traits typifying a different group according to its habitat and habitus, or ecological niche.

If this were so, then the process of evolution occurred over time—geological time, eons upon millennia, of which Lyell's *Principles of Geology* spoke. And, if it were so, there ought to be intermediates, transitional forms, species caught in the act of evolving from one thing to another. Where was this "infinitude of interconnecting links"?[9] Darwin mulled over this stumbling block for a long time, worrying at it like a dog at a favorite bone. Eventually he came to believe that most of these transitional species or populations would have been exterminated, because the linking populations would be by definition less well adapted than their adaptive neighbors, whose superior numbers and more suitable adaptations would overwhelm the intermediates. It was to be expected that the links would be missing, Darwin concluded contentedly. Once the intergradations or links were destroyed, what was left formed a diverse array of separate species, all sufficiently similar to be grouped into a single genus and each beautifully adapted to its habitat. Time and natural selection are the only two prerequisites.

"I am convinced," he wrote in *The Origin*, "that species are not immutable; but that those belonging to what are called the same genera are lineal descendents of some other and generally extinct species, in the same manner as the acknowledged variations of any one species are the descendents of that species. Furthermore, I am convinced that Natural Selection has been the most important, but not the exclusive, means of modification."

It does not sound like such explosive stuff, but it was and Darwin knew it. "I see no good reason," he continued in vain, in the second edition, "why the views given here should shock the religious feelings of any one."[10] But he did see a good reason—he was murdering belief as Macbeth had murdered sleep—and he was thus tacitly entreating his audience for clemency and moderation in their responses.

The crux of the problem was a single sentence, a sentence whose implication was so roundly, so fully, so bulgingly pregnant that it could not be overlooked: "Much light will be shed on the origin of Man and his history."[11]

That animals evolved, rather than being created by God fully perfect from the beginning, was bad enough. It disrupted the notion of things being pretty much as they always had been. The essence of evolution was change—Darwin's descent with modification—and change was not a desirable commodity in the complacent world of Victorian England. But this more stunningly simple idea, that humans, too, evolved, that humans, too, were part of nature, marked Darwin's work as the opening skirmish of a war. It could not have been readily foreseen then that this same war was to be fought on shifting battlegrounds until today without signs of abating.

What started the war was the astonishing power of Darwin's evolutionary paradigm to make sense, to bring a sort of ordered rationale to the chaos of life. Darwin's idea provided for many a blinding glimpse of the obvious. It *explained* what could be observed everywhere by those with keen enough eyes to see.

Life, Darwin's ideas optimistically implied, was ever-improving, ever-perfecting. Natural selection was a sort of progress machine, bettering everything in every way in a wonderful echo of Voltaire's character, Dr. Pangloss, who maintained cheerfully: "In this best of all possible worlds . . . all is for the best."[12]

Yet in the theory's power lay its destructiveness. Darwin's hypothesis did brutal violence to the world as it existed in England before 1859. It slaughtered long-held views of human specialness and distinctness from the brutish creatures of the world; it allied us with all life-forms, implying a continuity and coherency that was simultaneously breathtaking and horrifying; it emphasized the physical and left the moral, the intellectual, and the spiritual tenuously grounded; it annihilated life as Darwin, and everyone else, had known it. It denied—or was perceived to deny—the teachings of the Christian Church, the unspoken underpinning of nineteenth-century English society. Paradise was lost, indeed.

Or perhaps Paradise had never been. Victorian society (like all others) was an inconsistent patchwork of beliefs and practices, especially as applied to humans. Servants and common laborers might be treated as subhuman, left to succumb to disease, despair, poverty, and abominable living conditions. Even the social approbation heaped on those who did good works and nursed the poor was based on the exercise of *charity* by moral and financial superiors, not upon the recipients' common humanity or human rights. The people of other nations—often erroneously seen as other *races*—were even more dubiously human.

This conflict between the sentimental ideal of humanity and the flinty reality was epitomized by a tragic event that became a nine days' horror in England. Just days before Darwin's words would initiate the massacre of Christian belief, an acquaintance of his from the *Beagle* voyage—a Tierra del Fuegian named Jemmy Button—engaged in a bloody slaughter of Christian missionaries.

The history of Jemmy Button and his compatriots Fuegia Basket, York Minster, and Boat Memory exemplified the nineteenth-century Englishman's view of the humanness (or lack thereof) of different races.[13] Jemmy Button and the others were members of the Yahgan or the Alacaluf, two hunting and gathering tribes that eked precarious livings out of the marine coast of Tierra del Fuego, the southernmost tip of South America. Encountered on a voyage of the *Beagle* previous to Darwin's, these four Indians were summarily removed from their families and homes and transported to England by Captain FitzRoy. FitzRoy had initially seized the two adult men, York Minster and Boat Memory, as hostages after numerous episodes of thievery and assault. Fuegia Basket, about nine years old, had been apparently abandoned by her parents, and Jemmy Button, a youth, came aboard voluntarily out of curiosity. FitzRoy soon decided to take them back to England as curiosities, a course of action unthinkable if they were sentient humans like himself. The decision to seize these people seems to have been on a par with picking up the skull of an exotic tribesman, so that a learned account of its anatomical peculiarities could be presented to a scientific society in London—a popular endeavor of the day, if one performed with little scientific rigor or method by those who traveled to wild places. The knowledge of human variability, like the acknowledgment of the variability in humanness, was at a primitive and disordered state indeed.

FitzRoy's hope was that, once these Indians spoke English and had seen the wonders of the civilized world, their return to their native environment might spark a lively and profitable trade. His high-handed kidnapping was neither unique nor shocking. One of the four Indians died of smallpox and the others learned only rudimentary English. Three years after their departure, Jemmy, Fuegia, and York returned to Tierra del Fuego on the *Beagle* with Darwin; they were accompanied by Richard Matthews, their tutor from England, who proposed to start a mission. When the *Beagle* returned to check on them nine days later, Matthews, horrified by the savagery of the people among whom he found himself, feared for his life and begged to be taken aboard again.

His perception of danger was probably right. Some twenty years later, in November of 1859, Jemmy and several hundred Yahgans brutally massacred a group of English missionaries during a hymn service.

The assault followed months of classic, cross-cultural misunderstand-ings and conflicts. The event was half a world away from England, ge-ographically, and centuries away in terms of behavior. When the tale of this dreadful carnage in South America reached the ears of a shocked and Victorian England, the moral was clear. Darwin might say that humans had evolved, but here was proof positive that such primi-tive creatures as Jemmy were barely human at all. The beast was in them yet.

For his part, Darwin had spent the intervening twenty years finding the courage to explicate his deadly and heretical ideas openly—and even then, he hid behind vague symptoms of ill-health, shunning pub-lic meetings or discussions where he would be sure to be challenged. That he had avoided bringing them out into the open as long as possi-ble cannot be challenged. The ideas were nascent when he came home from the voyage of the *Beagle* in 1836 and "opened" a notebook on "the species problem" in 1837. He set about constructing a theory based on what he had seen, which was largely outlined by 1839. By 1844, he had talked and written of these ideas endlessly to friends; he had even sketched them out fairly clearly in an essay sent to Gray and, belatedly, to Hooker.

Given his peevish complaints of illness and discomfort—recorded in innumerable letters and a daily *Diary of Health* which he kept for five and a half years—it is tempting to attribute the delay to physical ill-health, but the facts show this is not an adequate explanation. It was the ideas themselves, the growing theory that, like a Medusa, turned him to stone and immobilized him for years.

When, after pondering the ways of creatures and their natural histo-ry, he had begun to glimpse the whole, the unifying design, it seemed at first too grand a vision, a scheme too large to grasp at once—an idea too complicated to explain. He could see it almost clearly enough to write it out, yet its ramifications haunted him and kept him awake at night.

Although his self-proclaimed major concern between 1839 and 1859 was with the origin of species, he was unable to complete a lucid and thorough explanation of his ideas. Yet somehow, during those years, he managed to write and publish innumerable papers and five major books on other, diverse topics. These were the journal of his *Beagle* voyage; a wholly new theory on the structure and making of coral reefs; two ambitious works, *Geology of Volcanic Islands* and *Geology of South America;* and an exhaustive monograph on barnacles. This prolifera-tion of other works reeks of avoidance and self-protectiveness.

Sensing the strife that lay before him, knowing his weaknesses more intimately than any other, Darwin sought to stockpile unimpeachable and overwhelming evidence. He burrowed into safe minutiae, keeping his eyes myopically focused on small details and facts. A perfect example was the barnacle study, an apparent detour from Darwin's main road that was explicitly undertaken to establish his credentials in taxonomy (the science of classification), in the hope that such an accomplishment might ease acceptance of his theory of the origin of species.[14]

The barnacle study came about like this. In 1845, Darwin had written to Hooker about a book on the species problem by Frederic Gerard, who specialized in botanical subjects. Hooker dismissed Gerard: "I am not inclined to take much for granted from anyone [who] treats the subject in his way and who does not know what it is to be a specific Naturalist himself."[15] Hooker bestowed far higher credibility on scientists who had carried out good, sound taxonomic studies.

The implied criticism hurt Darwin acutely, for he had no such studies to his name. He had been trying for some time to persuade Hooker to put aside his own concerns long enough to read the 1844 sketch on the origin of species. Darwin, ever self-doubting, may have suspected that Hooker could not find the time because he believed Darwin's ideas would be neither valid nor interesting. In fact, Hooker had previously referred to Lamarck's ideas on the mutability of species as "twaddle,"[16] so the likelihood of a fond reception for Darwin's thesis seemed remote. Was Hooker trying to postpone dealing with Darwin's ideas indefinitely, so as to avoid having to criticize his kind and eager friend? Perhaps. In any case, Darwin was quick to feel the cut himself when Hooker slashed at others for presuming to tackle the broad species problem without sufficiently detailed knowledge of any single group of organisms.

"How painfully (to me) true is your remark that no one has hardly a right to examine the question of species who has not minutely described many [himself]," Darwin wrote wretchedly, vowing to undertake such a study immediately. Hooker protested that his remarks were not meant to apply to Darwin, with his considerable experience as a naturalist. But Darwin knew a wound when it bled: "All which you so kindly say about my species work does not alter one iota my long self-acknowledged presumption in accumulating facts and speculations on the subject of variation, without my having worked out my due share of the species." He added one further guilt-inducing and self-denigrating remark: "But now for nine years it has been anyhow the greatest amusement to me."[17]

A month later, Darwin was at last almost poised to get on with his species book. "I hope," he wrote to Hooker, "this next summer to finish my S. American geology; and then to get out a little zoology and hurrah for my species work, in which according to every law and probability, I shall stick and be confounded in the mud."[18] But the "little zoology"—the barnacle study—proved a circuitous diversion far more slowing than the predicted mud.

Starting in 1846, Darwin spent eight years on barnacles. It became such a monumental, all-absorbing task that one of Darwin's children was heard to ask about a friend's father, "But where does Mr ———— do his barnacles?"[19] Ironically, after Darwin had worked for four years on this "little zoology"—four years marked by serious depression and even worse than usual health—Hooker finally got around to reading Darwin's 1844 sketch of his species theory. He wrote to ask about progress on that front rather than about barnacles, about which he had heard a great deal lately. Darwin chided him in reply: "By the way, you say in your letter that you care more for my species work than for the Barnacles; now this is too bad of you, for I declare your decided approval of my plain Barnacle work over theoretic species work, had very great influence in deciding me to go on with the former and defer my species-paper."[20]

Consumed though he was, Darwin was thoroughly tired of barnacles before the clinging creatures quit his life. He wrote to his cousin William Darwin Fox in 1852, even as the first volumes of his monograph were being published, "I hate a Barnacle as no man ever did before, not even a sailor in a slow moving ship."[21] Yet the barnacle volumes won him acclaim as a scientist of substance and a very welcome medal from the Royal Society. This accomplished, it might be supposed that the slowest-moving ship of all, Darwin himself, might now set full sail in a high wind for the species problem. It was still not to be.

For years more, Darwin simply talked and talked about his species theory, apologizing for its unclearness or unfinished nature. This habit of going on about his work, without ever articulating exactly what he meant or presenting the coherent whole, must have been trying for even his closest friends and confidants, some of whom had been hearing about Darwin's poor theory for more than fifteen years. Of course, Darwin's enthusiastic charm, his kindliness, and his wonderfully childlike delight in new facts or ideas largely mitigated the exasperation others may have felt at his slowness.

Still, even the great geologist Charles Lyell expressed some impatience with Darwin's reluctance to *get on with it*. After a visit to Down

House in 1856, Lyell wrote, "I wish you would publish some small fragment of your data, *pigeons* if you please and so out with the theory and let it take date and be cited and understood."[22]

In the end, Darwin's hesitations were overcome by an unpredictable turn of events involving Alfred Russel Wallace, a self-taught naturalist whose life was as shaped by deprivation as Darwin's was by comfort.[23] Where Darwin had wasted much of a privileged education, Wallace had learned hungrily during the few years of schooling his family had barely managed to afford. Apprenticed at fourteen, Wallace worked long hours at a series of poorly paid jobs, scraping together a few shillings to buy a book here or attend a lecture there. He was, to a large extent, an autodidact who would never enjoy the luxury of Darwin's university experiences: being tutored by Adam Sedgwick, the renowned geologist, or walking and talking with J. S. Henslow, the Cambridge botanist.

After years of saving and self-education, in 1848 Wallace finally expended his entire life's savings—100 pounds (an order of magnitude less than Darwin's annual private income)—to finance a voyage to the Amazon on the H.M.S. *Mischief.* Wallace and his friend Henry Bates planned to cover the rest of their expenses and perhaps turn a profit by collecting specimens for museums at home and abroad. Tragically, their adventures and trials came to little as the boat sank upon the return voyage. Wallace was left with only a few bedraggled items he had salvaged, those specimens that he had already shipped home, and some insurance money, for his agent had providentially insured the collection.

Undaunted, Wallace published what he could and soon embarked on another natural historical voyage, this time to the Malay Archipelago and southeast Asia. It was there, lying in a malarial fever, that he was suddenly struck with the notion of survival of the fittest as the mechanism behind adaptation and evolution. He had, like Darwin, been influenced by reading Malthus, who pointed out that each species produced far more offspring than could possibly live. What Wallace saw, then, was that the fittest survived—that natural selection would benefit those with advantageous adaptations and would ruthlessly prune out those who lacked them. Thus could species change; thus could evolution occur. Wallace wrote out his ideas in a feverish hurry, on two successive evenings, and decided to send them to Darwin, with whom he had an on-again-off-again correspondence. Darwin had been kind to him, treating him with respect and friendliness despite Wallace's distance from the closely woven fabric of the British scientific community.

Wallace little realized that he was sending a long-fused letter bomb, one that would explode Darwin's procrastination. His missive took four long months to reach Down House from Ternate, in the Moluccas, and included a manuscript entitled "On the Tendency of Varieties to Depart Indefinitely from the Original Type." Upon opening the communication in June of 1858, Darwin's genteel complacency and always precarious sense of well-being were shattered.

Far from home—and far from the educated and well-to-do circles of London's exclusive scientific societies even when he *was* at home— Wallace had innocently sent the paper to Darwin for his opinion, begging him to forward it to Lyell (one of the most powerful and eminent scientists of the day) if Darwin thought it good enough. Reading Wallace's lucid prose, Darwin came to the terrible realization that this man had seen the same vision, had arrived at the very same theory that lurked in Darwin's mind—from which he had been turning his head— for all those years. Wallace's previous paper, published in 1855, had shown he was treading close on Darwin's heels, although few in London scientific circles had paid Wallace's work much heed, perhaps because he was a relative unknown and certainly an outsider to them. And now it had occurred: another man had articulated the very precepts that Darwin had for so long held close to his heart.

Arriving in June, 1858, Wallace's new paper provoked a wild disorder in Darwin's mind—a clashing of rivalry, conscience, jealousy, honor, pride, self-criticism, selfishness and selflessness that left Darwin ill, weak, and utterly desperate. Compounding the problem was a genuine crisis of health in the household: not Darwin's, for once, but his children's. In mid-June, a horrifying epidemic of scarlet fever struck and threatened all of his beloved offspring. Two became ill and within the month the youngest, a retarded boy, was dead. By the end of the summer, so were five other children in the village and one of the Darwins' nurserymaids. In a family rife with hypochondria and practiced in translating psychological pains into physical ones, no more devastating combination than truly deadly illness and absolute scientific crisis can be imagined.

As ever, he turned to his friends. For his entire life, Darwin was one who sought constant reassurance and approval from others. His letters practically pleaded for attention and praise, and yet—to spare himself the awful possibility of rejection—he prefaced many of his discussions with self-deprecating remarks like those that perhaps his notions were "bosh"[24] that might "explode like an empty puff-ball."[25] In this, the worst calamity he had ever faced in his entire scientific career, Darwin threw himself into Hooker's and Lyell's hands in a series of letters that paint a clear image of his state of turmoil.

To Lyell, he wrote on June 18, 1858: "Some year or so ago you recommended me to read a paper by Wallace. . . . He has to-day sent me the enclosed, and asked me to forward it to you. . . . Your words have come true with a vengeance—that I should be forestalled."

Having broached the dreadful topic, Darwin now anxiously reminded Lyell of his (Darwin's) own claim to the ideas in Wallace's paper. He continued: "You said this, when I explained to you here very briefly my views of 'Natural Selection' depending on the struggle for existence. I never saw a more striking coincidence; if Wallace had read my MS. sketch written out in 1842, he could not have made a better short abstract!" Here is an interesting slip of the pen, for Darwin's sketch was written in 1844 and was nearly always referred to as such by him (including a remark in another letter, a week later, to Lyell). It seems likely that his intense concern for priority caused him to alter his normal usage in this first, urgent letter.

He went on, "Please return me . . . [Wallace's] MS., which he does not say he wishes me to publish, but I shall of course, at once write and offer to send to any journal." He concluded tremulously, "So all my originality, whatever it may amount to, will be smashed, though my book, if it will ever have any value"—an unkind critic at this point might have inserted *if it is ever written*—"will not be deteriorated; as all the labour consists in the application of the theory."[26]

After a week, Darwin wrote again, beseeching his friend to find him an honorable way out of the moral dilemma. Again, he reminded Lyell of his (Darwin's) priority. Again, he swung from despair—and threats to burn his book, which did not actually exist in tangible form—to hope. "I should be extremely glad now to publish a sketch of my general view in about a dozen pages or so," he wrote, though two years earlier he had rejected this very advice from Lyell. "[But] I cannot tell whether to publish now would not be base and paltry." He asked Lyell to send the letter, and his answer, on to Hooker, "for then I shall have the opinion of my two best and kindest friends. . . . This is a trumpery affair. . . . I will never trouble you or Hooker on the subject again."[27]

But Darwin could not leave the matter to rest there, try as he might. He wrote Lyell again the next day, balancing against his own claims his awareness of Wallace's potential indignation—for Wallace might well feel, if Darwin's ideas were published now, that Darwin had taken advantage of his openness and goodwill. Still, Darwin reasoned, "It seems hard on me that I should be thus compelled to lose my priority of many years' standing, but I cannot feel at all sure that this alters the justice of the case. First impressions are generally right, and I at first thought it would be dishonourable in me now to publish."[28]

To Hooker, on June 29, he wrote a final and most revealing letter. It started: "I have just read your letter and see you want the papers at once. I am quite prostrate and can do nothing, but I send Wallace, and the abstract of my letter to Asa Gray. . . . I dare say it is all too late. I hardly care about it."

This last statement is patently untrue; it was surely included to make Hooker feel less guilty if he disappointed Darwin by telling him that he could not publish under the circumstances. Darwin continued, reiterating the bonds of friendship and mutual respect: "But you are too generous to sacrifice so much time and kindness. It is most generous, most kind. I send my sketch of 1844 solely that you may see by your own handwriting that you did read it."

Again, Darwin's insecurity is manifest. Hooker was badgered by Darwin for years during the 1840s to read the sketch, and was most interested in it when he did. Thus Hooker was hardly likely to have forgotten the document—or to accuse Darwin of having just written it up with a false date.

In the closing lines, Darwin's mood fluctuates like a roller coaster, swinging wildly from anguish to hopeful ambition. No wonder Darwin's stomach churned.

> I really cannot bear to look at it. Do not waste much time. It is miserable in me to care at all about priority.
>   The table of contents will show what it is.
>   I would make a similar, but shorter and more accurate sketch for the "Linnean Journal."
>   I will do anything [you suggest]. God bless you, my dear kind friend.
>   I can write no more.[29]

Lyell and Hooker, responding both to Darwin's desperation and to their sense of morality, proposed to present both Wallace's paper and Darwin's 1844 sketch, plus a letter written in 1857 by Darwin to Asa Gray, to a meeting of the Linnean Society. Darwin was delighted; the deed was done in record time on July 1, immediately after Darwin's tortured letter reached Hooker. Whether in an attempt to secure his priority, or in simple obeisance to alphabetical order, Darwin's paper appeared first.

It was a perfect solution, at least from the points of view of the participants. Wallace was delighted to have his ideas presented to so prominent a body, with the clear sponsorship and protection of such eminent scientists as Lyell and Hooker. That his paper appeared jointly with Darwin's seemed of little concern to him. Darwin was relieved to recapture his priority without the moral stigma of having deliberately scooped Wallace. Hooker and Lyell were comfortable at having effected a Solomon-like judgment that neatly saved their friend from either despair or dishonor. And, to be fair, they may well have suspected that, once his priority was firmly established, Darwin would never finish his big book at all. In this case, Wallace would be able to claim the meat of the theory by producing a fuller treatment of it in short order. It seemed hardly likely that the tortoise, who had been plodding toward the species problem for twenty years, would beat the hare, who had dashed off a completed paper in two days.

The reception of the joint papers was curious and artfully managed. Hooker and Lyell made prefatory remarks explaining the awkward situation and the reason for the joint presentation to the Linnean Society. Then the papers were read (neither author being present). Afterward, they also commented "chiefly with a view of impressing on those present the necessity of giving most careful consideration to what they had heard," according to Francis Darwin, Charles's son and editor of his letters, who adds: "There was, however, no semblance of a discussion."[30]

As Hooker described the meeting to Francis Darwin:

The interest excited was intense, but the subject was too novel and too ominous for the old school to enter the lists [i.e., to sign up as military recruits], before armouring. After the meeting it was talked over with bated breath: Lyell's approval and perhaps in a small way mine, as his lieutenant in the affair, rather overawed the Fellows, who would otherwise have flown out against the doctrine. We had, too, the vantage ground of being familiar with the authors and their theme.[31]

It was a brilliant maneuver, well executed by Darwin's cronies. But their blessing could only stifle the immediate public reaction; they could not hope to suppress it altogether. In England, science, especially natural history, was so informal and so little organized (there were almost no professorships or curatorships formally designated as being for natural history) that even well-recognized giants like Hooker and Lyell did not have the power to demand adherence to their ideas. Thus, it was an era in which debate and disagreement flourished, with happy long-term consequences, despite the prestige and dignity of some of the participants. Of course, the unhierarchical nature of English science did not obviate the class system, which was part of why Wallace, teetering on the edge of the working and middle classes, would never achieve the fame or acceptance of the well-born Darwin.

Now, at last, Darwin began to work on his book. He had but narrowly escaped the threat of being "punished"[32] for his vanity and pride by being deprived of recognition for his many years of work on the species question. He started with what he continued to call the "Abstract" of his species book even when it swelled to 400–500 pages. Astonishingly, Darwin was able to muster his thoughts and put them down coherently with incredible speed. His diary notes that he started on the Abstract on July 20, 1858, worked on it for three weeks, took a break of a little over a month, and recommenced. A mere eight months after he started—by the end of March—Darwin was ready to send the manuscript to John Murray, the publisher.

A master at working the system, Darwin had arranged for Lyell's kindly intervention to smooth the road. Lyell spoke to Murray about the book even before Darwin mailed it off. Typically, Darwin's misgivings at the book's reception were never-ending. He wrote anxiously to Lyell:

> Would you advise me to tell Murray that my book is not more *un*-orthodox than the subject makes inevitable. That I do not discuss the origin of man. That I do not bring in any discussion about Genesis, &c. &c., and only give facts, and such conclusions from them as seem to me fair.
>
> Or had I better say *nothing* to Murray, and assume that he cannot object to this much unorthodoxy, which in fact is not more than any Geological Treatise which runs slap counter to Genesis.[33]

This last line is a clever reminder that Lyell's own great work, *Principles of Geology*, had also been perceived as anti-Biblical. But in the twenty-odd years that had intervened since Lyell's youthful publication and Darwin's magnum opus, Lyell had become far more conservative about such matters. Though he aided Darwin at many turns, he long hesitated to endorse Darwin's evolutionary ideas openly and in toto. They joked about the dire threat of Lyell becoming "perverted" by Darwin's ideas, though in fact Darwin was hurt at his friend's withheld imprimatur.

Murray agreed to publish principally on Lyell's recommendation, after seeing only the chapter headings. Darwin continued to fuss and worry, replying: "Accept your offer. But I feel bound for your sake and my own to say in clearest terms that if after looking over part of the MS. you do not think it likely to have a remunerative sale I completely and explicitly free you from your offer."[34]

In a later letter to Murray, Darwin mingles apology and pride:

> I send by this post, the Title (with some remarks on a separate page), and the first three chapters. If you have the patience to read all Chapter I, I honestly think you will have a fair notion of the interest of the whole book. It may be conceit, but I believe the subject will interest the public, and I am sure that the views are original. If you think otherwise, I must repeat my request that you will freely reject my work; and though I shall be a little disappointed, I shall be in no way injured.[35]

In view of his anguish over the Linnean affair, Darwin's last sentence can be read only as politeness rather than truth.

Once the proofs came, Darwin revised and rewrote the manuscript heavily, for which he apologized profusely to Murray. The book itself appeared on November 24, 1859, only a year and four months after the Linnean Society meeting. Darwin took good care to send prepublication copies to a long list of scientists in England, France, and the United States whose opinion was of especial concern to him. Each received a personal letter, begging humbly for fair consideration, careful reading, and, if they could spare the time, any lengthy or detailed criticisms or suggestions they might have about the ideas would be particularly valuable.

It was an excellent way to see that his book received the attention of all those who mattered; indeed, it was much talked of even before it was officially published. The entire printing of 1250 copies sold out "to the trade" on the day of publication, to Darwin's immense satisfaction. It also caused him considerable consternation. His health broken down by stress, Darwin was "under water-cure [hydropathy], with all nervous power directed to the skin, I cannot possibly do head-work, and I must make only actually necessary corrections."[36]

The long argument was started. Questions rang in the air like bullets, too dangerous to be ignored, too numerous to be dodged: Who was human and what does "human" mean? Where did we come from and who are "we"? How did we get here and where are we going? They pierce and wound us still.

And so, evolutionary theory started as a simple enough intuition, rooted in the unfocused observations of a well-to-do, spottily educated young man shocked by his first exposure to a world beyond the well-ordered borders of nineteenth-century England. It was a harsh and wonderful collision that fractured the tough seed and rooted the germ of a great idea in Darwin's mind. Spindly and weak, the concept of evolution gained strength slowly, taking twenty years to grow to a recognizable form. The idea slowly budded into a conviction that would bloom, whether its inadvertent gardener willed it or not, in 1859 with the publication of Darwin's *The Origin of Species*.

Evolution was no welcome idea planted with intent, whose glorious petals were awaited with eager anticipation. It was to Darwin a dark and cursed flower whose pungent fragrance stank of disorder, contention and, most headily, of change. But flower it did.

"How EXTREMELY STUPID not to have thought of that!"[1]

Thomas Henry Huxley found Darwin's theory astonishingly self-evident when he read *The Origin of Species*, though he had much disliked previous attempts at evolutionary theories and had soundly criticized Robert Chambers's *Vestiges of Creation* some years before. But others, especially those with strong, orthodox Christian beliefs, would engage in a more passionate struggle with Darwin's ideas.

Darwin had mused slowly for years, withdrawing step by step from society to the quiet of Down House, until Alfred Russel Wallace unwittingly provoked the greatest crisis of Darwin's life. Even a well-to-do, retiring, hypochondriacal gentleman of science like Darwin, prematurely elderly in manner but with an intelligence as keen as any young man's, could not bear the thought of losing his theory to another. It was Darwin's child, bastard though it might be, and he would not see Wallace put his name on it. Jealousy and pride swept away the qualms that had made Darwin hesitate for twenty years and, in less than a year, he finished *The Origin of Species*.

But publication alone was not sufficient. Though Huxley had been left out of the well-intentioned and largely innocent conspiracy over Darwin's and Wallace's Linnean Society papers, he proved to be the man of the hour. The mild-tempered "father" of evolutionary theory was himself hopelessly unable to defend his surprising creation from the heated battering of objections and criticisms that would try its strength, as were Lyell and Hooker. Lyell was secure enough in position and an unimpeachable public figure, but was unwilling for reasons of religion to throw himself into the fray. Hooker was not constitutionally suited to the task, either. But Huxley—quick-witted, irreverent, able, combative, unconventional, and a deucedly clever speaker—*Huxley* was the one Darwin needed.

Darwin knew it and so did Huxley. They recognized their mutual need and contracted without discussion an alliance that would ensure the survival and growth of Darwin's evolutionary theory. They were strangely complementary.

"Darwin," Huxley wrote in 1851, "might be anything if he had good health."[2]

Huxley had health but lacked another essential: "If he had leisure like you and me . . . ," Lyell wrote to Darwin, "what a position he would occupy!"[3]

And so, one with leisure, the other with health, they became together something more than they were separately. They were an odd couple joined in a peculiar sort of marriage, but it had an essential inevitability about it that could not be denied. "If I can convert Huxley I shall be content,"[4] Darwin wrote Hooker, shortly after he had bundled up a copy of his book with a typically self-denigrating letter and sent them to Huxley.

For Huxley, the book was like "the flash of light which, to a man who has lost himself on a dark night, suddenly reveals a road which, whether it takes him straight home or not, certainly goes his way."[5] It was the grand theory Huxley had been seeking, even if he had been only vaguely conscious of his search; when he read Darwin's words, he knew what he had found and knew what his role must be.

It had been a long dark road that led Huxley to this collaboration with Darwin. Huxley's personality, his intellect—his whole life in a sense—had been but preparation for their partnership, for, more than anything, Huxley was a man who loved to turn theories upside-down.

Perhaps it was a simple fascination with rigorous mental gymnastics; perhaps it was the sheer fun of the unexpected or his sense of delight at the revelation of the truth. Whatever the source, Huxley always enjoyed a surprising idea and never shied away from a good argument, as long as it was sound and well-grounded. If inversion showed up someone's pet idea as feeble—even if it were his own—so much the better. His success was built on cleverness, not name or family or fortune, which may have innoculated him against that endemic disease of the successful, pomposity.

He was born quick-witted and dark-eyed, both heritages from his

beloved mother. "I cannot help it," she had proclaimed characteristically. "Things flash over me."[6] And so they did over her son; for his
whole life, he was quicker, more perceptive, more brilliant than most
of those around him.

Intelligence was not a universally appreciated quality. As a boy, for
two unhappy years, Huxley attended the Great Ealing School—a well-
regarded public school—where his father was senior assistant master.
His memories of it were harsh. It was the era in which English public
schools followed the new ideas of Thomas Arnold at Rugby, designed
to build self-reliance and character. Living conditions were Spartan
and uncomfortable; a strong prefect system prevailed and covered up
much bullying and abuse of the younger boys. What was prized was ex-
cellence at team sports; what was sought were good, sound chaps, lacking emotionality and introspection and not too long on curiosity.

Huxley was not only the master's son, an awkward position, but also
a boy who in no way fit the desired description. From an early age, his
thirst for learning was insatiable, his desire to understand, keen, and
his urge to know more and learn faster than his fellows, aggressive. It
was a recipe for unhappiness in a British public school of that era, and
so it proved. From 1835 (when Huxley was ten and his father left Ealing to become a banker), Huxley suffered little more formal education.

For a different child, this change would have marked a sharp decline
of learning. For Huxley, the freedom of being able to read and explore
on his own—at last unfettered by pedantic, unimaginative teachers and
untutored classmates with unawakened intellects and dubious abilities—was a welcome release. He read voraciously—about chemistry,
physics, history, anatomy, natural history, geology, and philosophy. He
taught himself languages (French, German, and Italian) and carried out
experiments. He went haymaking with a hay fork in one hand and a
book in the other. He grew tall and slender, with a striking square jaw,
a pleasant countenance, and a handsome mass of thick, dark hair.

Though he had far fewer years of schooling than one might expect,
Huxley suffered from no want of learning and intellectual exercise. Two
experiences were to prove pivotal, points of inflection in the graph of
his life.

One was Huxley's abandonment of Christianity as a religion or system of beliefs—but not as a body of moral precepts or thoroughly Victorian emphasis on truth-telling, honor, and rectitude, which he upheld
throughout his life. The Huxley family, including Thomas, dutifully attended church services and, as a small boy, the local rector was one of
his prime heroes. He recounts, in his *Autobiography,* putting his

pinafore on backward to resemble clerical garb and delivering an earnest sermon in the rector's style to the giggling kitchen maids. But slowly this devotion turned to disbelief, to skepticism. The Church's emphasis on faith, on belief unsupported by evidence, seemed untenable to Huxley's growing confidence in intellect and reason. By his mid-teens, Huxley no longer saw himself as a Christian; later, he coined the usage "agnosticism" for his beliefs. He later explained:

> I, the man without a rag of a label to cover himself with, could not fail to have some of the uneasy feelings which must have beset the historical fox when, after leaving the trap in which his tail remained, he presented himself to his normally elongated companions. So I took thought, and invented what I conceived to be the appropriate title of "Agnostic." It came into my head as suggestively antithetic to the "Gnostic" of Church history, who professed to know so very much about the very things of which I was ignorant.[7]

In time, this confidence in the ability of the mind to grapple with reality—in contrast to the blind illogic of religious faith, which demanded belief without facts—became a trait so integral to Huxley's personality that it was passed down as a family attribute, along with a certain unconventionality. It is nowhere more clearly demonstrated than in the lives of his children. Many years later, Huxley's daughter Ethel—known as "The Dragon" to one generation and "The Grand-Dragon" to the next—was the family matriarch to whom plans and marriages were more or less submitted for approval. Taking aside the prospective fiancee of one grandson, she queried the girl sternly. "You are marrying into one of the great atheist families. I know you are an atheist now; but will you be able to keep it up until you die?"[8]

That Huxley from an early age was unencumbered by orthodox Christian beliefs would prove crucial to his acceptance of Darwin's theory. Most obviously, Huxley was spared the moral dilemmas and deep religious conflicts with which many scientists of the day struggled. How could a believer in the Biblical creation abandon this cornerstone of Victorian society in favor of Darwin's undirected, self-creating *natural* selection, however powerful the comparison to humanly guided selection might seem? More subtly, however, Huxley's early rejection of Church doctrine, coupled with his complete acceptance of the prescribed, day-to-day behavior of a Christian gentleman of the era, showed Huxley's intellectual independence, his willingness to test and weigh up any and all beliefs—however sacred—on the scales of intellect. This, too, would prove important in the role he was to play.

At about the age at which Huxley was separating from his religious beliefs, another event occurred that was of deep significance to him. Human anatomy and medicine were among the myriad subjects of interest to him, not least because two of his sisters had married physicians and such topics must have come up in conversation frequently. Thus it was that at thirteen or fourteen, Huxley attended a postmortem examination, or autopsy. He overcame the repulsion of the odors and the vivid grotesquery of cutting into a dead human body and stayed, fascinated, for hours to watch the physical and intellectual exposition of the workings of the body. He may have dissected actively, rather than simply observing; his *Autobiography* is vague on the point.

Dissection of a human cadaver is an experience that unnerves and deeply disturbs virtually all who observe it for the first time; this profound unease is the source of the crude and seemingly uncaring "black humor" for which modern medical students in the anatomy laboratory are noted. There is a profound emotional impact in a demonstration of mortality and human frailty, coupled with intense intellectual stimulation. It is an experience that no one ever forgets and one that frequently gives rise to nightmares and moral conflicts among modern students.

But medical students today are years older than Huxley, who was then in his adolescence and found the experience traumatic. He fell into "a strange state of apathy," an illness Huxley believed to be the physical result of poisoning. This is not an improbable hypothesis in the days before embalming of cadavers, but psychological shock was certainly a contributing factor. Desperate to cure his son of what appeared to be a rapid decline toward death itself, his father sent Huxley to the quiet of the Warwickshire countryside to friends. Huxley later recalled:

> I remember staggering from my bed to the window on the bright spring morning after my arrival, and throwing open the casement. Life seemed to come back on the wings of the breeze, and to this day the faint odour of wood-smoke, like that which floats across the farmyard in the early morning, is as good to me as the "sweet south upon a bed of violets."[9]

He had decided to live.

Though inclined toward mechanical engineering, Huxley's fascination with how things worked was sublimated once again into medicine, the profession of his two brothers-in-law. At sixteen, Huxley was apprenticed to a Dr. Chandler, whose patients were the miserably poor inhabitants of the London dock area; though the Huxley family was in somewhat straitened circumstances, Chandler's practice exposed Huxley to squalor, ignorance and poverty on a horrifying scale that he had never before experienced and never forgot. After some months he moved to Regent's Park, as apprentice to the physician husband of his favorite sister, Eliza Salt; this afforded him more time to study.

With only two years of formal schooling behind him, he began to consider enrolling at London University, to which end Huxley began attending a series of lectures on botany.

> One morning [he recounted in later life] I observed a notice stuck up—a notice of a public competition for medals etc. to take place on the 1st of August (if I recollect right). It was then the end of May or thereabouts. I remember looking longingly at the notice, and someone said to me, "Why don't you go in and try for it?" I laughed at the idea, for I was very young, and my knowledge somewhat of the vaguest. Nevertheless I mentioned the matter to [John Salt] when I returned home. He likewise advised me to try, and so I determined I would.[10]

He won the Silver Medal of the Apothecaries' Society for botany and a Free Scholarship to the Charing Cross Hospital Medical School, which paid the fees his family could never have afforded. It was the first of many medals and prizes Huxley was to win during his training, for he had natural ability and a tremendous power to concentrate and work very hard at subjects that interested him. He spent so much time bent over a microscope at the window that his friends teased him that his silhouette was the emblem of a fictional pub: the Sign of the Head and Microscope.[11] Looking back on his career, Huxley mused:

> I am now occasionally horrified to think how very little I ever knew or cared about medicine as the art of healing. The only part of my professional course which really and deeply interested me was physiology, which is the mechanical engineering of living machines . . . what I cared for was the architectural and engineering part of the business, the working out the wonderful unity of plan in the thousands and thousands of diverse living constructions, and the modifications of similar apparatuses to serve diverse ends.[12]

It was a new and fruitful way of looking at organisms. At twenty, Huxley published his first scientific paper, gathered up the prizes in chemistry, anatomy, and physiology, and graduated from medical school to face the prospect of earning a living. There was no question of buying into a practice nor did research positions, scarce as they were, pay much. Therefore, though he lacked the family contacts and connections that usually facilitated such arrangements, Huxley determined to enter the Royal Navy as a medical officer—and dared to write directly to Sir William Burnett, Director-General of the naval medical services, citing his credentials and requesting a position. He was granted an interview, passed an exam before the College of Surgeons, and duly became Lieutenant Huxley under the command of Sir John Richardson, the Arctic explorer, naturalist, and surgeon. It was Richardson who obtained for Huxley the post of assistant surgeon on the H.M.S. *Rattlesnake* for a four-year voyage to Australia, New Guinea, and the Pacific islands. Part of Huxley's duties were to assist ship's naturalist John MacGillvray in the collection and study of natural history specimens.

And, as he wrote happily to his sister Eliza:

[W]e shall form one grand collection of specimens and deposit it in the British Museum or some other public place, and this main object always being kept in view, we are at liberty to collect and work for ourselves as we please. Depend upon it, unless some sudden attack of laziness supervenes, such an opportunity shall not slip unused out of my hands.[13]

Like Darwin's voyage on the *Beagle* fifteen years earlier, Huxley's experience on the *Rattlesnake* was to form the foundation of his career. From notes in his diary, it is clear that Huxley formed an explicit plan about how to profit from this opportunity. Studies toward a systematic end—that is, in order to classify organisms taxonomically—were neither to his taste nor likely to be easily carried out without reference collections from other parts of the world. Taxonomy was better left to those ensconced in museums to do. "But what I *can* and they *cannot*," he added happily, "and where therefore the chief value of my position: I can observe 1. the 'habits' of living bodies, 2. their mode of development and generation, 3. their anatomy by dissection of fresh specimens, 4. their histology by microscopic observation."[14]

His working conditions were challenging, to say the least. The five-foot-eleven Huxley had a cramped cabin on a lower deck that measured seven feet by six feet, with a height of only four feet ten inches. He lashed a microscope to a bench to dissect specimens in poor light, sometimes with ankle-deep water sloshing underfoot. He and MacGillvray even had to improvise collecting-nets, though making a collection was explicitly part of the ship's mission.

Their requests for reference works had been politely ignored by the Admiralty, so the only scientific books at hand were his prized copies of the works of the noted French naturalist the Comte de Buffon. This led the sailors to refer to his specimens (when they were not tossing them overboard as being smelly and in the way), with cheerful tolerance, as "buffons." For pleasure, he had an Italian dictionary, a copy of Dante in Italian, Carlyle's *Sartor Resartus* and *Miscellanies*; these "were among the few books devoured partly by myself and partly by the mighty hordes of cockroaches in my cabin," he wrote.[15] Food was simple and monotonous, the biscuits were weevily, and days were long.

He loved it, as Darwin had loved his voyage on the *Beagle*. Up to this point, Huxley's life had differed strongly from Darwin's, but perhaps the coincidence of a wonderful voyage when young gave Huxley and Darwin a sort of shared experience that lent closeness to their later association.

Huxley adjusted to his cramped conditions, learned to enjoy his messmates, was enchanted by the new creatures and places, and grew into an artist/scientist who treasured specimens as much for their natural beauty as for their scientific interest. He dissected and drew exquisitely detailed renderings of the delicate sea creatures the nets brought in; he thought long and hard about their habits and anatomies, able to rely only on his wits, his eyes, and whatever knowledge he and MacGillvray had stored up in their heads.

Before the voyage was out, he began submitting scientific papers by mail to London journals, though there was a disheartening lack of response—apparently due only to the *Rattlesnake*'s irregular schedule. Upon his return to England he found two articles were already published and a third under consideration by the Linnean Society. His scientific career was well under way, though he did not know it yet. More important, he had met and fallen in love with Henrietta Heathorn, a charming young woman whose family lived in the outback 90 miles from Sydney; she would become his wife and lifelong companion.

Huxley was young, darkly handsome, self-confident, romantic, full of fun, and had no good prospect of being able to support a wife and family. In 1847, the two became engaged after only four meetings; in the next few years, they saw each other over a period of perhaps six

months, as the *Rattlesnake* explored the region and periodically returned to Sydney. In the end, the patient Henrietta would wait seven
long years before she and her "Hal" could be wed. Against the odds of
uncertainty, long separation, and brief acquaintance, it was to prove a
highly successful partnership. The Huxley household, full of children,
cats, art, and books, was called by friends the "Happy Family"—a reference to a children's game.

On his return to London in 1850, Huxley lobbied the Admiralty for
paid leave in order to complete his work and a grant to pay for publishing it. They granted the leave, at half-pay, but would promise no assistance with publication costs. At first, it seemed that Huxley's efforts
must be crowned with success. In amazingly short order, he was elected to the Royal Society and nominated for their Medal (which he would
not win until the following year, at the precocious age of twenty-seven).
Such recognition was heady stuff for a young man, and Huxley throve
on the fellowship, the praise, the intellectual stimulation of the gentlemen who made up the scientific societies of London—gentlemen
like the botanist Joseph Hooker, the physicist John Tyndall, the philosopher Herbert Spencer, and, of course, Charles Darwin, whom he val-
ued greatly. Huxley developed his writing skills and a witty yet concise
style, soon finding that he could earn more by writing than through his
naval pay. He was not yet making enough money to send for Henrietta
and marry.

In January 1854, Huxley was still awaiting either a favorable decision on publishing his *Rattlesnake* works or promotion, inducements
that had been held out to him upon enlistment. His request for further
leave provoked orders to join the *Illustrious* at Portsmouth and, tacitly, to abandon his attempts at publication. Despairing of his situation,
Huxley turned again to the unconventional tactic that had served him
well before: he wrote directly to the Secretary of the Admiralty pointing out that neither the promised promotion nor the assistance with
publication had been forthcoming. "Considering the distinct pledge
given in the minute [of his original interview] to which I have referred,
then granting it would seem as nearly to concern their Lordships' honour as my advantage."[16]

It was an audacious and blunt reproach that no upper-class gentleman (the usual type of officer) would have issued. The response was a
month of thunderous silence followed by a terse command to join the
Illustrious, as previously ordered, or face separation from the Navy.
Believing himself to be in the right, Huxley refused to join the new ship
until a clear decision on the question of a grant for publication was

made. The Navy had no interest in making a clear decision and soon divested itself of this awkward young man. Huxley was now in worse circumstances than ever, save that he was free to take up another position, could one be secured.

He wrote some articles, gave some public lectures, and simply held on, hoping for change. Indeed, once he had developed the knack of public speaking—a task he once regarded as an awful chore—he became known as one of the most brilliant and clear-minded speakers on scientific subjects of his day. This, or time, seemed to transform his fortunes. First came a chance to deliver a series of lectures at the Royal (or Government) School of Mines, leading to a steady lectureship that grew into a professorship of natural history. Then a part-time appointment was offered at the Geological Survey and then another, at the department of comparative anatomy at St. Thomas' Hospital. Huxley's reputation, earned solely through hard work and brilliance and owing not one whit to family connections, spread. Finally, the Royal Society, unwilling to support publication by a naval officer, joined forces with the Ray Society to pay for publication of Huxley's *Rattlesnake* works. In 1855 he was at last able to marry.

Huxley's lectures and articles are a model of logical clarity, of simple observations transformed into illustrative examples of deep significance. And yet, there was always the playfulness, the willingness to laugh at himself, the jest of a big-hearted man lurking behind the mask of utter respectability and seriousness. It is no wonder that Huxley, with this combination of clear thinking and good humor, was able to break into the circles of the amateur gentlemen-scientists generally closed to those from good but middle-class families.

And in 1855, Huxley also embarked upon a surprising (for any other man of science of the era) series of lectures: scientific lectures aimed at the working man, supposed by most to be incurious, inarticulate, and congenitally limited in intellect. He was convinced that common people could and should understand science—that scientific understanding must illuminate ordinary lives. And the ordinary people of London were no more immune to Huxley's "glorious, sublime"[17] lectures than were students or fellow academics. He spoke to standing-room-only audiences of 600, in lecture theaters filled with mechanics, cabbies, journeymen of all sorts, and common laborers. He developed the rare talent of walking the tightrope that refused to compromise either truth or clarity. He became a sort of celebrity, on one occasion finding a cab driver returning his fare with these words: "Oh, no, Professor. I have had too much pleasure and profit from hearing you lecture to take any money from your pocket—proud to have driven you, sir."[18]

It was not always thus, however. Huxley himself—ever one to deflate self-congratulatory pride, even his own—told the story of one of his early speaking engagements:

> In my early period as a lecturer I had very little confidence in my general powers, but one thing I prided myself upon was clearness. I was once talking of the brain before a large mixed audience and soon began to feel that no one in the room understood me. Finally I saw the thoroughly interested face of a woman auditor and took consolation in delivering the remainder of the lecture directly to her. At the close, my feeling as to her interest was confirmed when she came up and asked if she might put one question upon a single point which she had not quite understood. "Certainly," I replied. "Now Professor," she said, "is the cerebellum inside or outside of the skull?"[19]

The abrupt deflation of the lecturer's pride on this occasion was a good reminder not to misjudge the audience—and Huxley never forgot this incident.

As a rising star of anatomy and paleontology, and one given to skewering obfuscating, long-winded, and self-important scientists (as well as himself), it was perhaps inevitable that Huxley should fall afoul of Richard Owen, the "British Cuvier." Georges Cuvier was the renowned genius of comparative anatomy in late eighteenth- and early nineteenth-century France—the man who could, proverbially, reconstruct an entire extinct animal from a single one of its bones because of his exhaustive command of anatomy. Had he still been alive, Cuvier might have quibbled with the application of this name to Owen; Owen himself certainly did, feeling no doubt that he (Owen) was the better man. Indeed, Owen felt himself superior to most "and does not conceal that he knows it," as Huxley wryly remarked after their first meeting.[20]

Though in temperament and style the two could not have differed more completely, there was an odd series of parallels between Huxley and Owen, ten years senior. Like Huxley, Owen was born to a middle-class family and, at sixteen, was apprenticed to a physician because medical school fees were beyond their means. Like Huxley, Owen's life was also much affected by dissection of cadavers. But Owen recounted no such delicacy or horror at the procedure as Huxley. Indeed, whereas Huxley's first reaction to the ghoulish, but essential, task of dissecting a human body was to fall peculiarly ill, Owen often spoke with fondness of his happy days as a teenager, carrying out postmortem dissections on the prisoners who died at Lancaster Gaol.

The task sounds horrific, combining the standard horrors of dissection (probably in crowded and none-too-clean quarters) with the distasteful handling of bodies that would almost invariably have been filthy, neglected and diseased. However, such autopsies would perhaps have been more palatable since the dregs of society confined to prisons were hardly considered human by most respectable Victorians; it may have been more akin, emotionally, to dissecting animals than fellow human beings. In any case, whatever the drawbacks, it was a splendid and almost unparalleled opportunity to learn. Ordinary medical students in the Victorian era had perhaps one chance a year to observe a dissection, unless they wished to resort to paying body-snatchers to provide contraband corpses.

The job provided Owen with a tremendous chance to learn and understand human anatomy; it even allowed him to satisfy his curatorial bent. Owen's biography includes a story from those days that apparently he often told as a party piece, under the name of "The Negro's Head":

My zeal and skill at assisting at *post-mortems* had gained me the rarely bestowed commendation of the doctor, our preceptor. I had already begun to form a small anatomical collection, and had lately added a human cranium to my series of dogs and cats and the skeletons of mice and such "small deer." It happened also that on the day when a negro patient in the gaol hospital had died, a treatise on the "Varieties of the Human Race" fell into my hands, and greatly increased my craniological longings. The examination of the body was over and the hurried inquest performed, when, slipping some silver into the hand of the old turnkey as we left the room, I told him I should have to call that evening to look a little further into the matter, before the coffin was finally screwed down. . . .

[After hours], provided with a strong brown-paper bag, I sallied forth . . . to secure my specimen of the Ethiopian race. I was now an *habitué* of the place, and an attendant was no longer proferred to accompany me. Taking my lantern and keys, I opened every door and gate, duly locking them again after I had passed through. . . . The gloom of the apartment was just made visible by the light of the lantern, but it served to the business immediately in hand. The various instruments had judiciously been left behind; and when I returned through the gates—the bag under my cloak—the intimation that all was now ready for interment was received with a nod of intelligence by the old turnkey. . . .

As soon as I was outside I began to hurry down the hill; but the pavement was coated with a thin sheet of ice, my foot slipped, and, being encumbered with my cloak, I lost my balance and fell forward with a shock which jerked the negro's head out of the bag, and sent it bounding down the slippery surface of the steep descent. As soon as I recovered my legs I raced desperately after it, but it was too late to arrest its progress. I saw it bounce against the door of a cottage facing the descent, which flew open and received me at the same time, as I was unable to stop my downward career. I heard shrieks, and saw the whisk of the garment of a female, who had rushed through an inner door; the room was empty; the ghastly head at my feet. I seized it and retreated, wrapping it in my cloak. I suppose I must have closed the door after me, but I never stopped till I reached the surgery.[21]

It is a macabre tale, well told and full of atmosphere, but one too lacking in sensitivity to endear. Owen's admitted lust for a skeletal specimen of another race only underscores the utter lack of scientific knowledge of human variability.

Owen grew to be a pompous man, tall but lacking in grace, with bulging, exophthalmic eyes and a strong chin. His speech was full of overlong words and circuitous phrases—designed more as a display of erudition than for communication—and his written style has been described as "at best stupefyingly dull, and at worst, incomprehensible."[22] He was overly sensitive as to his status and position, perhaps because of having risen from the middle class through hard work.

Yet Owen was, without doubt, prolific and he enjoyed a widespread reputation as an anatomist and paleontologist of great ability. He transformed the Hunterian Museum at the Royal College of Surgeons—long a muddle of unlabeled pickle jars and anonymous bones—into "the most beautifully planned and the most conveniently arranged museum perhaps in Europe."[23] He moved to the British Museum in 1856.

Unfortunately, Owen was not a likable man. Few of his colleagues considered him a friend and many flatly loathed him. He was, in fact, widely known for his malicious backbiting, for his free usurpation of others' ideas, and for his jealous ambition. Even Darwin, a gentle, kindly man always inclined to give others the benefit of the doubt, disliked and avoided him, describing Owen harshly in his autobiography.

The same year that Owen took up his post at the British Museum, he sought and received permission to use the lecture hall at the Royal School of Mines (Huxley's institution) for a course in paleontology. It was a curious move, because Owen would presumably have had access to halls at either the Royal College of Surgeons or the British Museum. The next year, Owen self-styled himself "Professor of Comparative Anatomy and Palaeontology, Government School of Mines" in Churchill's *Medical Directory*. This was a deliberate insult to Huxley, who, as Professor of Natural History at the School of Mines, was responsible for those very topics while Owen had no appointment at the school whatsoever.

It was the first thrust in an academic duel which Huxley pursued with glee henceforth, delighting in showing up fallacies and inaccuracies in, for example, Owen's analysis of the vertebrate skull as a structure that was no more than an elaborated vertebra or his spurious identification of anatomical features that separated humans from apes. The enmity between Huxley and Owen would spread to other topics, too, such as Darwin's new theory of evolution.

If Huxley supported evolutionary theory, then Owen must be against it. Huxley enlisted in Darwin's army knowing what lay ahead. He wrote Darwin:

> I am prepared to go to the stake, if requisite, in support of Chapter IX [on the imperfection of the geological record as an explanation for the lack of transitional forms], and most parts of Chapters X, XI, XII [on the geological succession of organic beings, and on the geographical distribution of living forms]; and Chapter XIII [on the affinities of organic beings as judged from their morphology, embryology, and rudimentary organs] contains much that is most admirable, but on one or two points I enter a *caveat* until I can see further into all sides of the question.
>
> As to the first four chapters [in which Darwin drew parallels between variation in domesticated animals and "under nature" and put forth his ideas on the struggle for existence, natural selection, and the survival of the fittest], I agree thoroughly and fully with the principles laid down in them. . . .
>
> I trust you will not allow yourself to be in any way disgusted or annoyed by the considerable abuse and misrepresentation which, unless I am greatly mistaken, is in store for you.
>
> Depend on it, you have earned the gratitude of all thoughtful men. And as to the curs which will bark and yelp, you must rec-

ollect that some of your friends, at any rate, are endowed with an
amount of combativeness which (though you have often and just-
ly rebuked it) may stand you in good stead.

I am sharpening up my beak and claws in readiness.[24]

Those who argue that Huxley expressed doubts about various aspects
of Darwin's theories are correct. Huxley had misgivings and reserva-
tions about the workings of natural selection, which he regarded as not
yet proven in a mode that was thoroughly satisfactory to him. As an em-
piricist, grounded in laboratory work, Huxley immediately saw that
"there is no positive evidence, at present, that any group of animals
has, by variation and selective breeding, given rise to another group
[i.e., a new species] which was, even in the slightest degree, infertile
with the first."[25]

Darwin had a somewhat different philosophy of science and, conse-
quently, other ideas about what constituted proof. For him,

> this hypothesis may be tested—and this seems to me the only fair
> and legitimate manner of considering the whole question—by try-
> ing whether it explains several large and independent classes of
> facts; such as the geological succession of organic beings, their
> distribution in past and present times, and their mutual affinities
> and homologies. If the principle of natural selection does explain
> these and other large bodies of facts, it ought to be received [as
> true].[26]

In other words, Huxley sought experimental evidence and Darwin
sought explanatory power and plausibility. Huxley wanted to *see* a new
species arise, experimentally; he wanted to breed for variations, select
rigorously, and demonstrate that two new species, infertile with each
other, could be produced. Darwin wanted to put together a hypothetical mechanism that would plausibly explain his diverse and puzzling
observations. Why did the rate of increase of animals always greatly
exceed the rate of survival? How did it happen that there were groups
of broadly similar-appearing species within single geographic regions?
What accounted for the slow and successive appearance of new fossil
species in the geological record? And how did the suitability or adaptedness of a creature to its habitat and life-style arise? Little wonder,
with such different premises about the nature of science, that Huxley
maintained a few caveats about Darwin's work.

Yet perhaps emphasizing their disagreement obscures the larger and more important point: Huxley staked his professional life on *The Origin of Species*. He might call for more proof and additional experimental evidence, but his crucial, public defense of evolution can leave no doubt about where his allegiance lay. Evolution was *his* theory, the one that made sense of his world and drew out the underlying rules of nature. Another's infant it might be, abandoned on Huxley's doorstep, but he willingly took it in. There can have been no more adoptive parent, nor one fiercer in protecting his charge, than Huxley.

Huxley's actions were bold, more than mere passive caretaking. He adopted the theory and raised it, showed what it could do, as Darwin never did. Indeed, it can be argued that not only Huxley's intellect but his career had been waiting, unconsciously, for a truly great theory upon which they could let loose.

Until he read *The Origin*, Huxley was a man of strong intellect and admirable abilities, but he had not yet found the cause sufficiently bold, the idea sufficiently important, to animate his life into something greater. His fortuitous partnership with Darwin made them both. And it made something more: their partnership gave evolution life, it secured the survival of this fragile yet powerful idea through the dangerous days of its infancy, and it turned the idea toward its greatest subject, mankind.

The Question of Questions for Mankind

CHAPTER 3

BOTH HUXLEY AND DARWIN knew the animosity and vehemence that Darwin's theory might—and did—arouse. There could be no mistaking that the most incendiary issue would be the evolution of humans, even though Darwin had all but omitted it from his manuscript. The status and value of humans was rapidly becoming a public issue in any case, as Europeans confronted the wickedness of the thriving slave trade in Africa and the awful, progressive extermination of the Aborigines in Australia—not to mention the near-slavery of the European convicts transported for life to this inhospitable land. Evolutionary theory could not escape becoming associated with radical and contentious ideas like those espoused by the Abolitionist movement or prison reformers. In the face of such urgent moral dilemmas, could Victorian England or the rest of the world be persuaded that science and rationality were more glorious and comforting than traditional religion? It was worth a try.

Darwin, in his own way, had fortified his defenses. He had spent twenty years gathering data, anecdote by anecdote and observation by observation, all of which seemed to confirm his thesis. Reticent and nonconfrontational, Darwin had tried as much as possible in the writing of the book to cloak his contentious offspring in a woolly thickness of easily accepted facts, for protection. He had lobbied, obliquely, all of his friends and many of his scientific acquaintances that they should not think poorly of him for endorsing the mutability of species and that they would share with him their valuable opinion of his poor creature.

It was not enough. Fortuitously, the man at *The Times* assigned to review *The Origin of Species*, a Mr. Lucas, was flatly and openly ignorant of matters biological—"as innocent of any knowledge of science as a babe" he called himself.[1] He begged Huxley to ghost-write the review for him, and Huxley accepted with glee.[2] Huxley wrote another review for *The Westminster Review*. In both cases, the veil of anonymity was so gauzy as to be transparent. Yet these were important preemptive strikes that gave Huxley the opportunity to back the book publicly and make it one that everyone must read.

> This most ingenious hypothesis [Huxley avowed] enables us to give a reason for many apparent anomalies in the distribution of living beings in time and space, and that it is not contradicted by the main phenomena of life and organization appears to us to be unquestionable; and, so far, it must be admitted to have an immense advantage over any of its predecessors.[3]

Then he rather cleverly delineated the grounds on which he thought the battle should be fought. What remained to be determined was neither the explanatory power of the theory nor the existence of the "modifying causes and selection power, which Mr. Darwin has satisfactorily shown to exist in Nature,"[4] but simply whether Darwin had overestimated the efficacy of natural selection. It was not a matter of "if"; it was a question of "how much."

The review, prominently placed and brilliantly worded, ensured that Darwin's ideas would be given a trial and not simply ignored to death: "And whatever they do," Huxley wrote in a letter to Hooker, "they *shall* respect Darwin."[5] To be sure, *The Origin* would probably have sparked a controversy even without Huxley's reviews. Darwin's practice of dispensing advance copies had spread the book throughout Britain, Europe, and America and made its subject matter well known in advance. The book first appeared in bookstores on November 24, 1859; the second printing, with corrections and additions, appeared January 7, 1860.

And the criticisms and objections lagged not far behind Huxley's praise. These were so rapid, and so pertinent in some cases, that several corrected and amended editions followed the second in short order. These were not always minor changes; Darwin struggled against the long-standing habit of shying away from naked statements of his beliefs in order to make himself clear, seeking a way to express his theme through, not in spite of, the myriad observations he included. He also tried also to answer criticisms. As Morse Peckham, a Darwin scholar, has observed, "Of the 3,878 sentences in the first edition, near-ly 3,000, about 75 per cent, were rewritten from one to five times each. Over 1,500 sentences were added, and of the original sentences plus these, nearly 325 were dropped."[6]

Predictably, Richard Owen—Huxley's old nemesis—wrote an es-pecially poisonous review of *The Origin* in the *Edinburgh Review* in April of 1860. In a single issue of the magazine, he managed to exco-riate Darwin's book, a lecture of Huxley's, and some work of Hooker's as well. To Lyell, one of the few of his close associates spared Owen's venom, Darwin wrote:

> I have just read the *Edinburgh [Review]* which is without doubt by Owen. It is extremely malignant, clever, and I fear will be very damaging. He is atrociously severe on Huxley's lecture and very bitter against Hooker. So we three *enjoyed* it together. Not that I really enjoyed it for it made me uncomfortable for one night; but I have got over it today. . . . It is painful to be hated in the intense degree with which Owen hates me.[7]

He added in a letter to Hooker on the same subject: "In simple truth I am become quite demoniacal about Owen—worse than Huxley. . . . yet I never shall forget his cordial shake of the hand, when he was writing as spitefully as he possibly could against me."[8]

What was it that Owen said that so disturbed Darwin? With sarcasm veiled as wit, Owen wrote:

> The scientific world has looked forward with great interest to the facts which Mr. Darwin might finally deem adequate to the sup-port of his theory on this supreme question in biology, and to the course of inductive original research which might issue in throw-ing light on that "mystery of mysteries" [i.e., the origin of species]. But having now cited the chief, if not the whole, of the original observations adduced by its author in the volume now before us, our disappointment may be conceived. Failing the ad-equacy of such observations, not merely to carry conviction, but to give a colour to the hypothesis, we were then left to confide in the superior grasp of the mind, strength of intellect, clearness and precision of thought and expression, which might raise one man so far above his contemporaries, as to enable him to discern in the common stock of facts, of coincidences, correlations and analogies in Natural History, deeper and truer conclusions than

his fellow-labourers had been able to reach.

These expectations, we must confess, received a check on pe-
rusing the [very] first sentence in the book.[9]

And on he proceeded, making a cruel mockery of Darwin's every
phrase, willfully misunderstanding the point of some to make them appear absurd and apparently deliberately misquoting others, to the same
effect. Even as the defenders thronged to Huxley's flag, the offensive
forces swarmed behind Owen's colors.

Owen's was certainly not the only critical review; far from it. Darwin's former Cambridge professor, Adam Sedgwick, was quite as firmly set against evolutionary ideas as Owen. He was perhaps more
opposed, having nurtured Darwin's intellect when he was at university—and, by the same principle, Darwin was more deeply hurt by Sedgwick's words.

> I cannot conclude [Sedgwick wrote in *The Spectator* in March,
> 1860] without expressing my detestation of the theory, because
> of its unflinching materialism;—because it has deserted the in-
> ductive track, the only track that leads to physical truth;—because
> it utterly repudiates final causes and thereby indicates a demor-
> alised understanding on the part of its advocates. In some rare in-
> stances it shows a wonderful credulity. . . .
>
> But any startling and (supposed) novel paradox, maintained
> very boldly and with something of imposing plausibility, pro-
> duces, in some minds, a kind of pleasing excitement, which pre-
> disposes them in its favour; and if they are unused to careful
> reflection, and averse to the labour of accurate investigation, they
> will be likely to conclude that what is (apparently) *original*, must
> be a production of original *genius*, and that any very much op-
> posed to prevailing notions must be a grand *discovery*,—in short,
> that whatever comes from "the bottom of the well" must be the
> "truth" supposed to be hidden there.[10]

Boldly maintained paradoxes, pleasing excitement, wonderful creduli-
ty, and lack of careful reflection sound not one whit like the middle-
aged Darwin of Down House who hesitated for twenty years to publish
his theory, though the youthful Darwin Sedgwick had known at Cambridge may have been more like this description. In fact, these words
might be seen as suggesting that a gullible Darwin has been hood-
winked by a sort of snake-oil salesman of dubious science.

In addition to florid reviews, such as these, *The Origin* provoked a basketful of letters to the newspapers, damning sermons (*The Origin* became a kind of "anti-Bible"[11]), theological protests from scientists and scientific protests from theologians, and—from a few of those who actually accepted Darwin's theory—claims that the idea had previously been suggested by others (usually the writer himself). Darwin was alternately damned as a heretic or a fool, congratulated as a proud father, or left bewildered by incomprehensible (to him) criticisms. He was the subject of newspaper cartoons and parodies, to the point that no mention could be made of monkeys or apes without invoking the name of Darwin as well; he was, in short, a public figure.

"Half the fools throughout Europe write to ask me the stupidest questions," he grumbled,[12] perhaps ironically aware that his own endless letters of inquiry while he was formulating *The Origin* might have struck some recipients similarly. It all eventually told on the state of his health, of course, so that his role in the fray was played henceforth from the sidelines at Down House, in absentia. He had given birth to the idea; now it was someone else's problem.

Thus it fell to Huxley to defend Darwin's ideas at the famous meeting of the British Association for the Advancement of Science at Oxford in June of 1860. The timing was perfect: just enough months had elapsed since publication for *The Origin*—or the ideas expressed in it—to be the controversial centerpiece of several of the sessions. It was now transparently clear that the crux of the fight, the pivot on which the armies turned and wheeled, was the evolution of humans and the theological implications of this fact.

The papers of Thursday, June 28, brought the issue out into the open. Dr. Charles Daubeny, an Oxford professor of botany, spoke on sexuality in plants, offering his lifetime of observations of botany and geology in support of Darwin's theories. In the discussion that followed, Huxley avoided comment, feeling the forum was not the right one. The President of the Association, Richard Owen, suffered from no such delicacy. He launched into an open criticism of *The Origin*, citing a specific anatomical structure, the hippocampus minor, that he asserted was present in human brains but absent from that of the gorilla. The hippocampus minor would thus prevent the one from being descended from the other. Huxley, knowing full well from his own research that Owen was wrong on this point, leaped to his feet to offer a "direct and unqualified contradiction"[13] to Owen's assertion, making clear that publications demonstrating the truth would be forthcoming. Though few in the audience knew or cared what the hippocampus minor was, Huxley's challenge to Owen, and Owen's disagreement with Darwin, were obvious.

The height of the battle would be reached the next day, with the speech to be delivered by Samuel Wilberforce, Bishop of Oxford and a clerical proxy for Owen's point of view. He was also the Church incarnate, in this instance. Wilberforce had already written a critical analysis of *The Origin of Species* for the *Quarterly Review*. Unlike Owen, however, Wilberforce was an effective and compelling speaker and a renowned public figure. Wilberforce was known by the sobriquet "Soapy Sam," a nickname derived from the quip of a Queen's Counsel, Sir Richard Bethell, about the judgment of bishops being "oily and saponacious [sic] and therefore difficult to grasp."[14] The Bishop was ready to demolish Darwin's theory during the Saturday afternoon session. It was apparently common knowledge that the excuse for his prepared speech would be a talk by an American, Dr. Draper, on "The Intellectual Development of Europe Considered with Reference to the Views of Mr. Darwin."

Huxley, tired of the meeting and longing for peace and quiet, was on the verge of leaving rather than face being "episcopally pounded"[15] by an audience sure to be heavily seeded with Owen supporters. But Robert Chambers, author of the woolly-headed and rather ill-fated *Vestiges of Creation*—a vague, evolutionary book that he had published in 1844—stopped Huxley in the street and chided him for "deserting." So Huxley stayed to attend the fateful session, along with so many observers that a sudden change of venue had to be organized to accommodate them.

Over 700—some say 1,000—laymen and women, students, clerics, and scientists crowded into the long west room of the University Museum, finding seats on chairs and windowsills or standing in the aisle as they could. Summer sunlight streamed in through the windows, illuminating the discussants seated on the stage. Caricatures that appeared in *Spy* soon after the event contrast the beefy, aging, red-faced bishop, dressed in his gorgeous clerical robes and snowy linen shirt, with the pale-faced, slender Huxley drawn in black frock-coat, with jet-black hair and pronounced sideburns, his pince-nez clamped on his nose.

Poor Draper droned on for an hour, little attended to, as everyone waited impatiently for the impending explosion. The heat and tension mounted; patience, especially of the boisterous undergraduates, wore thin. Finally, Draper's interminable paper was over and the floor opened for discussion. Three men spoke—or attempted to—and were shouted down within the first nine minutes. Then at last the Bishop responded to calls for his remarks and began to pontificate.

It was an eloquent speech, persuasive yet largely lacking in scientific substance, and one well calculated to win his audience. Its very vagueness made it difficult to attack and Huxley sat listening and waiting for some glimmer of an opportunity, some solid shred of argument upon which to base a reply. Contemporary accounts of Wilberforce's address differ somewhat, but all agree that he made a fatal error late in his presentation. Turning toward Huxley, who was seated on the stage next to Sir Benjamin Brodie, president of the Royal Society, Wilberforce passed a sly and unforgivable remark: "I should like to ask Professor Huxley, who is sitting by me, and is about to tear me to pieces when I have sat down, as to his belief in being descended from an ape. Is it on his grandfather's or his grandmother's side that the ape ancestry comes in?"[16]

Slapping his hand on his knee, Huxley turned to Brodie and exclaimed, "The Lord hath delivered him into my hands!"[17]

Wilberforce's gibe transgressed an ironclad Victorian convention. No gentleman would cast unfounded aspersions on a lady's reputation—and a lady Huxley's grandmother must surely have been. It was a vulgar and ill-bred comment at best, especially coming from a clergyman. For similar reasons, no lady's name was ever brought into "business" discussions. This notion of cloistering ladies from the day-to-day world was expressed most clearly in the military prohibition against mentioning a lady's name in the mess; to do so not only invoked a fine but also brought immediate disapproval from one's colleagues. Wilberforce, perhaps lulled by the fine flow of his oratory on that hot afternoon, had shot himself in the foot by mentioning Huxley's grandmother.

Huxley arose immediately after the Bishop stopped speaking. John Richard Green, an undergraduate present at the fray, recounted in a letter Huxley's stunnning closing remarks:

> I asserted, and I repeat, that a man has no reason to be ashamed of having an ape for his grandfather. If there were an ancestor of whom I would feel shame in recalling, it would be a *man*, a man of restless and versatile intellect, who, not content with an equivocal success in his own sphere of activity, plunges into scientific questions with which he has not real acquaintance, only to ob-

scure them by an aimless rhetoric, and distract the attention of his hearers from the real point at issue by eloquent digressions, and skilled appeals to religious prejudice.[18]

Huxley had ignored (or refused to reply to) the slur on his grandmother entirely. There was a moment of dazed silence and then the listeners trumpeted their approval with applause and laughter.

The event was all but over. A lady listener, "employing an idiom now lost,"[19] expressed the tension and portent of the moment by fainting and being carried out of the room. Hooker, somehow afraid that Huxley's deadly quiet delivery had not carried to the entire crowd, rose and spoke bluntly of Wilberforce's utter ignorance of the rudiments of botanical science and patent unfamiliarity with Darwin's book. Admiral FitzRoy, Darwin's erstwhile captain of the H.M.S. *Beagle,* rose, waving a copy of *The Origin*, and confessed his pain that Darwin had published such an antireligious work. He avowed he had often reproached Darwin for possessing views antithetical to the Biblical Genesis. But there was no question, in newspaper, private, or verbal accounts, that Huxley had won the day for Darwin.

Darwin, characteristically, was ill and dispirited at home. But a letter from Hooker describing the grand finale of the British Association meeting cheered him wonderfully.

> I had been very poorly, with almost continuous bad headache for forty-eight hours [Darwin wrote back to Hooker], and I was low enough, and thinking what a useless burthen I was to myself and all others, when your letter came . . . your kindness and affection brought tears into my eyes. Talk of fame, honour, pleasure, wealth, are all dirt compared to affection. . . . I am astonished at your success and audacity. . . . I have read lately so many hostile views, that I was beginning to think that perhaps I was wholly in the wrong and that Owen was right when he said the whole subject would be forgotten in ten years; but now that I hear you and Huxley will fight publicly (which I am sure I never could do), I fully believe that our cause will in the long run prevail.[20]

How well Darwin knew himself and understood his debts to others. To Huxley, he chortled, "How durst you attack a live bishop in that fashion? I am quite ashamed of you! Have you no respect for fine lawn sleeves? By Jove, you seem to have done it well."[21]

The skirmishes continued and only later was it clear that the Oxford meeting was the decisive battle. In the scientific journals, there were wonderfully esoteric thrusts and counterthrusts about the hippocampus minor and other details of human and ape anatomy that meant little to nonspecialists. (In fact, the hippocampus minor controversy was to reach such lengths—and to sound so ridiculous to the layman—that it was satirized by Charles Kingsley in his book *Water Babies* as the "hippopotamus major" which was said to be found within the heads of men.)

But perhaps the most interesting development was that Huxley, with his usual penchant for educating the common man, entered into another series of lectures in 1860; the subject was "The Relation of Man to Lower Animals." The lectures—and the small book of essays derived from the lectures, *Evidence as to Man's Place in Nature*, which was published in 1863—showed what could be done with Darwin's theory by someone of Huxley's talents. Huxley's words went straight to the heart of the matter—no endless anecdotes about six-toed cats or bumblebees or pigeon-breeding with oblique implications, no coy "Light will be thrown on the origin of man and his history."[22] No. He took up the most controversial and the most important issue, facing it squarely and openly in front of crowds of hundreds of listeners who were perfectly capable of being excruciatingly rude if they thought a lecturer was speaking rubbish.

Evidence as to Man's Place in Nature is a gem of a book. "What is the good of my writing a thundering big book [*The Origin of Species*], when everything is in this green little book so despicable for its size," Darwin wrote Huxley,[23] jokingly but a trifle peevishly, after reading it. "In the name of all that is good and bad, I may as well shut up shop altogether." Indeed, if the book is an accurate recounting of the lectures, the series must have been stunning. It was certainly popular.

"My working-men stick by me wonderfully," Huxley wrote to his wife, showing his awareness of the progressive decline in attendance at lecture series if each presentation fails to be clear and engaging, "the house being fuller than ever last night. By next Friday evening they will all be convinced that they are monkeys."[24]

His 1862 lectures on the same subject in Edinburgh—a stronghold of clerical opposition to evolution—produced such enthusiasm that the audience was said to be as exuberant as a football crowd after a victory over England. But those who failed to attend were shocked and horrified. The *Witness*, an Edinburgh newspaper, was outraged: why had the Edinburgh Institute asked Professor Huxley—"the advocate of the vilest and beastliest paradox ever vented in ancient or modern times amongst Pagans or Christians"—to speak in public "and exhibit his anti-scriptural and most debasing theory of the origin and kindred of man"? Given the audience's favorable response to Huxley's words, the author was amazed that, at the end of the lecture, "the hearers refrained from forming themselves into a 'Gorilla Emancipation Society,' and from forming some prompt measures for humanizing and civilizing their unfortunate brothers."[25]

Indeed, when Huxley came to publish the lectures, the topic was still incendiary. As he later reminisced:

"Magna est veritas et pravalebit!" Truth is great, certainly, but, considering her greatness, it is curious what a long time she is apt to take about prevailing. When, towards the end of 1862, I had finished writing "Man's Place in Nature," I could say with a good conscience, that my conclusions "had not been formed hastily or enunciated crudely." I thought I had earned the right to publish them and even fancied I might be thanked, rather than reproved, for so doing. However, in my anxiety to promulgate nothing erroneous, I asked a highly competent anatomist and very good friend of mine [Lyell] to look through my proofs and, if he could, point out any errors of fact. I was well pleased when he returned them without criticism on that score, but my satisfaction was speedily dashed by the very earnest warning, as to the consequences of publication, which my friend's interest in my welfare led him to give. But as I have confessed elsewhere, when I was a young man, there was just a little—a mere *soupçon*—in my composition of that tenacity of purpose which has another name; and I felt sure that all the evil things prophesied would not be so painful to me as the giving up that which I had resolved to do, upon grounds which I conceived to be right. So the book came out; and I must do my friend the justice to say that his forecast was completely justified. The Boreas of criticism blew his hardest blasts of misrepresentation and ridicule for some years; and I was even as one of the wicked. Indeed, it surprises me, at times, to think how any one who had sunk so low could since have emerged into, at any rate, relative respectability. [Since then, the idea] has had the honour of being freely utilized without acknowledgment, by writers of repute; and finally it achieved the fate, which is the euthanasia of a scientific work, of being inclosed among the rubble of the foundations of later knowledge and forgotten.[26]

What was it Huxley said that produced such emotion, positive or negative? In the first essay, Huxley dealt systematically with the natural history of the apes—chimpanzees, gorillas, orangutans, and gibbons—from earliest accounts, writen by explorers, to Paul Du Chaillu's more recent, spine-chilling stories about savage gorillas. Running throughout the essay are striking comparisons of apes to humans, in gait, anatomy, expression, and other behaviors. Thus Huxley established the great similarity between humans and apes as a foundation for later arguments that (as in human families) close resemblances bespoke common descent.

The second essay, "On the Relations of Man to the Lower Animals," forms the heart of the book. In it, Huxley addressed what he calls

> The question of questions for mankind—the problem which underlies all others and is more deeply interesting than any other . . . the ascertainment of the place which Man occupies in nature and of his relations to the universe of things. Whence our race has come; what are the limits of our power over nature; and of nature's power over us; to what goal are we tending; [these] are the problems which present themselves anew and with undiminished interest to every man born in the world.[27]

He then proceeds to marshal the evidence from embryology and comparative anatomy that, bit by bit, leads the reader (or listener) to the inevitable conclusion that humans and all other mammals share a common design: "identical in the physical processes by which he originates—identical in the early stages of his formation—identical in the mode of his nutrition before and after birth . . . [and] a marvellous likeness of organization."[28]

Humans are but one point on a continuum of creatures, Huxley argued. If we were "scientific Saturnians . . . discussing the relations [of the animals of the Earth] to a new and singular 'erect and featherless biped,' . . . brought . . . for our inspection, well preserved, may be, in a cask of rum,"[29] we should at once place the new specimen within the category of placental mammals and would unerringly rank it closest to the apes.

Huxley then embarked on a tour-de-force summary of the anatomy of the primates, from lemurs to humans, from brain structure to toe bones. He took the discussion of the brain as an excellent excuse for a long and detailed footnote in which he sniped at Owen's erroneous statement that the human brain differed in structure from that of the apes—another salvo in the hippocampus debate. But Huxley omitted this footnote from the 1900 reprinted version of the book, on the grounds that "the verdict of science has long been pronounced upon the questions at issue"[30]—and decided, it might be added, in Huxley's favor. A modern scholar of Huxley's work, Mario di Gregorio, has pointed out that, when written, the book was "more against Owen than for Darwin,"[31] so that the excision of the footnote altered markedly the apparent significance and intention of the book.

While there can be little doubt that besting Owen lent particular zest to Huxley's treatment of the topic, it is also obvious that it *was* a Darwinian manifesto. Huxley doubtless hoped the reader (or listener) would walk away thinking Owen a fool and Huxley a most knowledgeable and astute anatomist. But he also hoped to convince his audience that they were all monkeys, as he wrote his wife—that they shared every intimate detail of their being with the nonhuman primates and that this commonality of structure was evidence of common ancestry. This was crux of the second essay.

> But if Man be separated by no greater structural barrier from the brutes than they are from one another [Huxley argued]—then it seems to follow that if any process of physical causation can be discovered by which the genera and families of ordinary animals have been produced, that process of causation is amply sufficient to account for the origin of Man. In other words, if it could be shown that the Marmosets, for example, have arisen by gradual modification of the ordinary Platyrrhini [New World monkeys], or that both Marmosets and Platyrrhini are modified ramifications of a primitive stock—then, there would be no rational ground for doubting that a man might have originated, in the one case, by the gradual modification of a man-like ape; or, in the other case, as a ramification of the same primitive stock as those apes.[32]

In the next paragraph, he brought up Darwin's hypothesis explicitly and, with disarming candor, admitted both its tremendous explanatory power ("Mr. Darwin has satisfactorily proved that what he terms selection, or selective modification, must occur, and does occur, in nature; and he has also proved to superfluity that such selection is competent to produce forms as distinct, structurally, as some genera even are"[33]) and his own wish for further, physiological experiments to prove Darwin's case.

This discussion is often cited as evidence of Huxley's ambivalence about Darwin's work. Yet it must be remembered that he openly endorsed Darwin's view: "I, for one, am fully convinced, that if not precisely true, that hypothesis is as near an approximation to the truth as, for example, the Copernican hypothesis was to the true theory of the planetary motions."[34] It was well within Huxley's intellectual abilities to defend, promulgate, and, in his own words, "adopt" Darwin's evolutionary theory while yet longing for additional evidence of a type more suited to his temperament and training. Indeed, his thoughtful hesitations and qualifications persuaded more compellingly than blind acceptance would have.

Anticipating his audience's response as they realized just how fully they had followed Huxley to the brink, Huxley closed with a brilliant argument. He first expressed their objection:

> On all sides I shall hear the cry—"We are men and women, not a mere better sort of apes, a little longer in the leg, more compact in the foot, and bigger in the brain than your brutal Chimpanzees and Gorillas. The power of good and evil—the pitiful tenderness of human affections, raise us out of all real fellowship with the brutes, however closely they may seem to approximate us."

Unable to resist another dig at Owen, Huxley then countered with his own response:

> To this I can only reply that the exclamation would be most just and would have my own entire sympathy, if it were only relevant. But, it is not I who seek to base Man's dignity upon his great toe, or insinuate that we are lost if an Ape has a hippocampus minor. On the contrary, I have done my best to sweep away this vanity. I have endeavored to show that no absolute structural line of demarcation, wider than that between the animals which immediately succeed us in the scale, can be drawn between the animal world and ourselves. . . . At the same time, no one is more strongly convinced than I am of the vastness of the gulf between civilised man and the brutes; or is more certain that whether *from* them or not, he is most assuredly not *of* them.[35]

In the final essay, Huxley conducted a similarly masterful review of the fossil evidence that bore on the question of human origins. Here, he had the advantage over Darwin. In 1859, when *Origin of Species* was published, there was no convincing evidence of the existence of fossil humans in the form of ancient beings, nonmodern in form, whose remains were preserved under conditions that attested to their antiquity. Such material had, indeed, been found—in 1856, in the Neander Thal (Neander Valley) in Germany, that subsequently lent its name to the type of fossil human whose skeleton was first recognized there. And although ferocious debates over the antiquity and normality of the Neandertal skeleton were raging in Germany almost immediately, these had not made their way into the Anglophone literature when Darwin was writing.

Of course, human antiquity had been long suspected and even supported by the discovery of ancient stone tools, sometimes associated with extinct, Ice Age mammals. But crudely worked stone tools—sometimes grouped with stone objects with such dubious signs of human intervention that they cannot be seriously credited as tools—are a different matter from skulls. The fossilized human skeletal remains that had so far been discovered and discussed were little different from those of modern humans. Thus, for English speakers, the first solid report of a fossil and *different* human came in April, 1861, when George Busk, an anatomist and friend of Huxley and Darwin, translated a paper by the German anatomist Hermann Schaaffhausen on the newly discovered Neandertal remains.

The upshot of Schaaffhausen's analysis was that the skeleton was genuinely ancient, had massive, robust limbs, and a large cranium with strongly projecting browridges like those found on apes. While some—notably a coterie closely associated with Richard Owen that included C. Carter Blake and James Hunt—questioned both the find's antiquity and its normality (a point much contested in Germany, where the remains had been branded pathological), Huxley and others were quick to understand exactly what this fossil meant.

Huxley sent for photographs and a plaster cast of the Neandertal skull so that he could compare it with another fossil cranium, from Engis in Belgium, that had been discovered in 1833. He worked systematically, using a standardized set of measurements, angles, and indices that were innovative at the time. The pertinent point was that the Engis skull, though associated with worked flints and bones of extinct animals, was no more or less than a "fair average human skull" in Huxley's assessment—he measured skulls of modern or recent humans from various races for comparison—while the Neandertal fossil was decidedly primitive and archaic in anatomy.

And indeed, though truly the most pithecoid [apelike] of known human skulls, the Neanderthal cranium is by no means as isolated as it appears to be at first, but forms, in reality, the extreme term of a series leading gradually from it to the highest and most developed of human crania.

. . . the fossil remains of Man hitherto discovered do not seem to me to take us appreciably nearer to that lower pithecoid form, by the modification of which he had, probably, become what he is. And considering what is now known of the most ancient Races of men; seeing that they fashioned flint axes and flint knives and bone-skewers, so much the same pattern as those fabricated by the lowest savages at the present day . . . we have every reason to believe the habits and modes of living of such people to have remained the same from the time of the Mammoth and the . . . Rhinoceros till now. [Therefore to find] the fossilized bones of an ape more anthropoid, or a Man more pithecoid . . . we must extend by long epochs the most liberal estimate that has yet been made of the antiquity of Man.[36]

Huxley used the review of fossil finds to argue, first, that human antiquity had been demonstrated; second, that the existence of a pithecoid or apelike human form more primitive than any yet known had been proven; and third, that fossils still more convincingly transitional between humans and apes were expected to be found in yet older geological strata. In sum, the series of essays (or lectures) showed through behavior, anatomy, embryology, and paleontology that humans were ruled by the same laws of nature as other creatures and were undeniably related to other primates, especially apes. Here was the evidence for *human* evolution which Darwin either did not know or dared not present in *The Origin of Species*.

Though Huxley's book was not universally admired, it was widely influential because of its clarity, brevity, and focus—three attributes rarely allotted to Darwin's book. *Man's Place in Nature* is often cited as the beginning of physical anthropology, which encompasses the study of the bodily variation and evolution of humans. Yet Huxley's book would have been robbed of significance, then as now, without the broader theoretical context already established by Darwin's.

Had it been a deliberate strategy—and it was not—the campaign could not have been better planned. Let Darwin's grander scope and wide-ranging ideas go first, drawing the fire and taking the brunt of the attack; let evolution then be defended by the quick-moving and quick-witted orator who distills the vague theory down to its essence, bringing to everyone's consciousness the fact that it is the evolution of *humans* that is the sticking point; and, finally, let the carefully ordered armies of pertinent evidence be mustered and engaged until it must be conceded that they have won the day. It was Huxley's book that proved anatomically and paleontologically that humans are, inextricably, part of nature and—more precisely—just another species of primate.

In England, by the end of the 1860s, the fight for the acceptance of evolutionary theory was over. This is not to say that evolutionary theory was wholeheartedly adopted as the prevailing paradigm; in many circles, that revolution would have to await the integration of Darwinian theory with modern genetics.[37] But the public cannonading had stopped and the bitter attacks had died down; witness that, in 1871, Darwin found the courage to publish *The Descent of Man and Selection in Relation to Sex*, wrestling with a subject he had previously abjured as too dangerous.

In England, in an odd way, evolutionary theory paved the road down which traveled the human fossils that were beginning to be discovered and recognized. Huxley had managed to protect Darwin's tender theory, working for science against religion, and it survived and flourished.

In the United States (when the fighting of the Civil War was not completely eclipsing all intellectual concerns) the Huxley role was taken by the American botanist Asa Gray, whose data on the distribution of plants had been used by Darwin and who had suggested printing an American edition of *The Origin*. Gray's primary antagonist was Louis Agassiz, the Swiss anatomist and geologist who had immigrated to the United States in 1846. But the scuffles between Gray and Agassiz had none of the verve or drama of the skirmishes in England. The pair were less flamboyant, less witty, or simply less committed to the fight. And the bitter discussions of the humanity, or inhumanity, of enslaved Africans were focused, sadly, as much on politicoeconomic issues as on the thin biological evidence.

The more interesting confrontation was shaping up in Germany, sparked by the translation of *The Origin* into German in 1860. On the one side was Huxley's good friend, the biologist Ernst Haeckel; on the other was Haeckel's onetime mentor, the pathologist Rudolf Virchow, also known as the pope or pasha of medicine. They were giants of their time, leaders of German science, and their enmity was implacable.

It was a caustic struggle that lasted for years between two men who contrasted strongly in physical type and personality: Haeckel was Germanically bluff, blond, and handsome, with a strong romantic streak and a love for the out-of-doors; Virchow, the consummate laboratory man, was small, dark, wiry, pedantically precise, and naturally acerbic. It seemed at first a simple matter, Haeckel's engaging openness in defense of evolutionary theory pitted against Virchow's controlling, skeptical repression of new ideas. Yet the circumstances were somehow bizarrely transmuted, until light proved to be darkness and reason was twisted into horrific distortion. In the end, the natural villain of the piece, Virchow, was fighting a deeply sinister devolution of Darwin's theory that began in the seemingly benign hands of Haeckel.

Part II
Evolution
Evolving

Enthusiastic Propagandism

CHAPTER 4

POWERFUL AS THEY WERE, Darwin's words and ideas could not travel of their own volition. They required a courier, a messenger to transmit them in accessible language to new minds, an eager gardener to make them the focal point of a new arrangement of ideas. In England, this mercurial role was fulfilled by Huxley, aided by Darwin's other friends and colleagues, such as Lyell and Hooker. In Germany, the main agent of transplantation—labeled by Francis Darwin, Charles's son, as the practitioner of "enthusiastic propagandism"[1]—was Ernst Haeckel.

But the initial presentation of *The Origin of Species* was by Heinrich Georg Bronn, a senior and well-respected paleontologist at the University of Heidelberg. He was a logical choice as translator. In 1857, Bronn won an international prize, the Grand Prix, awarded by the Académie des Sciences in Paris for an essay on the laws of development of the natural world: more or less the same topic as Darwin's work. Bronn's expanded essay was published in German in 1858, two years before his translation of Darwin's book appeared. Unlike Darwin, however, Bronn was a professional scientist—neither Germany nor France enjoyed the widespread existence of amateur gentleman scientists that typified the English scene—at the top of his career in a hierarchical and rigidly structured community.

Bronn's view of reality was based, in part, on a lifetime of meticulous compilation of the animal remains found as fossils in different strata, or geological layers. He found that the faunas, or suites of species that were entombed together, varied from layer to layer, but to his eyes changes occurred in a gradual or progressive way, rather than abruptly or suddenly. The theory of catastrophism—the notion that life on earth was repeatedly decimated by massive geological catastrophes that swept continents clean, only to be refurbished with inhabitants by a new Creation—found little support in his studies. Since he rejected the catastrophist interpretation that had been first put forth by the French anatomist Georges Cuvier, Bronn devised his own theory, relying on the empirical evidence of the fossils themselves and taking as his motto *Natura doceri*[2] ("to be taught by Nature" or "let Nature teach").

But Bronn was not immune to theoretical influences. In fact, like many other German naturalists of the era, Bronn's ideas were strongly colored by the tradition of Goethe and his *Naturphilosophie*. This was a Romantic school of thought that postulated that the structure of organisms reflected an archetype or master plan devised by a higher being. *Naturphilosophie* stressed, in a mystical sense, the oneness of God and Nature. For Bronn, this philosophy was expressed in his theory of the history of life as the continuous development of God's underlying plan or intention. Although species were fixed or unchanging, according to Bronn, the range of species at any one point in time reflected a maturing plan analogous to the development of an organism from birth to adulthood. Later organisms were more advanced and represented progress over earlier ones of similar design. Humans were, of course, the end-point of this developmental evolution.

It was also a rule of nature that organisms needed an adaptedness, or suitability, to the environments in which they found themselves. Over time, the creative force thus produced new and genuinely novel species, cunningly adapted to different habits and lives and seemingly each more perfect and more intricate than that which preceded it— a manifestation of the exquisite unfolding of a master (and almost certainly Divine) plan.

For Bronn, evolution was explicitly analogous to the development of the organism, with a preordained pathway from embryo to mature form;[3] this idea lent distinctly teleological overtones to the same sorts of observations that Darwin had made from a different perspective. Bronn's was a common point of view and similar, developmentally-based readings of evolution were to gain considerable ascendancy late in the nineteenth century. Thus, although Bronn was a competent and interested translator, he was by no means a particularly sympathetic or unbiased one.

Perhaps it was his lack of endorsement of Darwinism that led him to make a critical change in his translation of *The Origin of Species*; he chose to omit Darwin's most pregnant sentence: "Light will be shed on the origin of Man and his history."[4] This apparently harmless sentence, for all its understatement and lack of aggrandizement, symbolized the crux of the issue in England. Relatively few cared at a deep emotional or intellectual level about the origin and history of, say, barnacles or primroses. What mattered to Darwin, Huxley, Soapy Sam Wilberforce, and thousands of laypeople was where humans belonged in the grand scheme of things and whether they could be accounted as more or less than, or simply different from, the beasts.

This was much more an abstract, theoretical concern; the distinction between humans and animals, which were perceived as symbols of the worst of human behavior and impulse, was taken very seriously. And yet, "human" was generally taken to mean "like us" in a class, cultural or ethnic sense rather than "any member of the human species."

It was simply not clear to nineteenth-century thinkers that all humans were in truth a single species. The pronounced physical differences among the different racial groups that were being encountered with some regularity by European explorers, missionaries, and colonists suggested otherwise. Even the ability of the different races to interbreed and produce healthy offspring was seriously debated.[5] These perceived physical differences were only heightened and reinforced by differences in dress, mores, manners, and beliefs. Those living among people of widely different cultural traditions were bewildered and sometimes defeated by the social and behavioral chasms that separated them. What was the *value* of the differences?

Darwin himself took Jemmy Button's tribe, one of the indigenous peoples of Tierra del Fuego, at the tip of South America, to be relicts, late-surviving remnants of ancestral, primitive humans. Like many others of his day and class, he simply did not believe that these savages ("absolutely naked and bedaubed with paint . . . their mouths, frothed with excitement . . . their expression . . . wild, startled and distrustful . . . like wild animals"[6]) were human in the same sense that he himself was. He was more horrified by their lives than he was impressed with their similarity to those among whom he had grown up. Nineteenth-century anthropologists were generally concerned with tracing the migrations of various human groups (Celts, Saxons, and so forth) around the continental Old World. This led to a certain obsession for acquiring and measuring skulls of various "races," which were in fact more often ethnic or tribal groups; ancient skulls, whether fossilized or simply old, were even more prized. But how to measure, what to measure, and what the significance of differences in those measurements *meant* was far from clear. Physical anthropology was in an infancy marked by the dominance of anecdote over measurement, by description over hypothesis-testing.

Whether simply preliterate or actually prehistoric, the peoples whose skulls were so earnestly perused in the first half of the nineteenth century were fully human and entirely modern anatomically. The polygenists, led in England by Richard Owen or in Germany by Rudolf Virchow, believed that the races had been created (or had arisen) separately and were thus as different as, say, robins and pelicans. The monogenists, including Huxley and Darwin, championed the common origin of all humans from a single apelike ancestor. Across Europe and England, the polygenist versus monogenist debate was a long and bitter one that persisted for decades, even after the possibility of evolution was widely admitted.

If it was not at all apparent to the educated middle and upper classes of the era what humanity or humanness consisted of, the attributes of animality, primitiveness, savagery, and brutishness were nonetheless well defined. Whether or not one of the inferior classes, races, or species could actually be humanized through education and training remained a moot point. Indeed, this was to some extent the point of Captain FitzRoy's transportation of Jemmy Button and three other Tierra del Fuegians to London and it is exactly the issue that George Bernard Shaw was later to satirize in *Pygmalion*, in which much is made of the astonishing transformation of a filthy, crude flower seller from the gutter into a beautiful, genteel young woman of society.

Bronn's translation obscured this link between *The Origin* and the issues of humanity and human evolution. Bronn even appended a lengthy exposition of his objections and criticisms of the work, generally undermining Darwin's credibility. However, the power and importance of Darwin's ideas for human evolution were not hidden from Ernst Haeckel, a rising young naturalist and anatomist.

Haeckel was twenty-six when he read *The Origin*, much younger and more flexible in his ideas than the aging Bronn. A strapping youth, Haeckel was broad-shouldered, blond, bearded, and handsome, with a deeply rooted love of the German countryside and outdoor sports such as climbing, hiking, and swimming. His bluff and vigorous Germanic manner would later prove a trial to the quiet, reserved Darwins when the younger man came to sit at the feet of his hero.

"Häckel [sic] came on Tuesday," Emma wrote her son Leonard. "He was very nice and hearty and affectionate, but he bellowed out his bad

English in such a voice that he nearly deafened us." On the occasion of that visit, there was also a tea party at the Darwins' house that included Haeckel and another German professor, with a deaf wife. Describing it, Emma sighed, ". . . anything like the noise they made I never heard. Both visits were short and F[ather] was glad to have seen them."[7] Emma herself, the letter implies, was happier to have seen the back of them, knowing full well what havoc loud voices and boisterous good humor would have on Darwin's health.

Why did Haeckel seize upon Darwin's evolutionary theory and not Bronn? Haeckel's receptivity lay perhaps in his youth and his specific experiences. Like Bronn, Haeckel was thoroughly imbued with the ideas of *Naturphilosophie*. In Haeckel's case, however, it took the form of a passionate boyhood hobby which gave an excellent excuse for long, rambling walks in the countryside combined with intellectual stimulation: collecting and classifying plants for his home herbarium. Haeckel's love of botany confronted him with a singular and practical difficulty, he recalled in later life:

The problem of the constancy of transmutation of species arrested me with a lively interest when, twenty years ago, as a boy of twelve years, I made a resolute but fruitless effort to determine and distinguish the "good and bad species" of blackberries, willows, roses, and thistles. I look back now with fond satisfaction on the concern and painful scepticism that stirred my youthful spirits as I wavered and hesitated (in the manner of most "good classifiers," as we called them) whether to admit only "good" specimens into my herbarium and reject the "bad," or to embrace the latter and form a complete chain of transitional forms between the "good species" that would make an end of all their "goodness." I got out of that difficulty at the time by a compromise that I can recommend to all classifiers. I made two collections. One, arranged on official lines, offered to the sympathetic observer all the species, in "typical" specimens, as radically distinct forms, each decked with its pretty label; the other was a private collection, only shown to one trusted friend, and contained only the rejected kinds that Goethe so happily called "the characterless or disorderly races, which we hardly dare ascribe to a species, as they lose themselve in infinite varieties," such as rubus, salix, verbascum, hieracium, rosas, cirsium, &c. In this a large number of specimens, arranged in a long series, illustrated the direct transition from one good species to another. They were the officially forbidden fruit of knowledge in which I took a secret boyish delight in my leisure hours.[8]

This pretty tale, perhaps embellished a little as the reminiscences of great men sometimes are, nonetheless demonstrates Haeckel's willingness to see species as both mutable and variable—two key elements in Darwin's hypothesis. While Bronn's life was spent dealing with dead, static fossil remains, neatly separated into geological compartments, Haeckel struggled with the living variability of nature. Too, Haeckel had read Darwin's account of the *Beagle* voyage as an impressionable child, indulging in hero-worship of this apparently glamorous traveling naturalist and vowing to embark on a similar trip when *he* became a man.

By the time that *The Origin* was available in Germany, Haeckel had undergone a considerable intellectual voyage if not such a lengthy physical one. Though he aspired to become a botanist, his lawyer father urged a more practical choice of profession and the young Haeckel dutifully enrolled in medical school at the University of Würzburg. Haeckel's letters home to his parents, with whom he enjoyed a warm and loving relationship, reveal his charming and enthusiastic character, as well as his youthful lack of focus. In one letter he sings the praises of his extraordinary professors—and a gifted group they were, since Würzburg was at the time the leading school of the new science of histology—and, in the next, he plummets from intellectual excitement to boredom, overwork, and a fervent desire to drop out of his courses.

Yet he benefitted immensely from this education in the long run. Würzburg boasted a faculty that was expanding the frontiers of knowledge, mapping millimeter by millimeter the cellular structure and function of the tissues of the body and their role in health and disease. Haeckel's teachers included Albert Kölliker, who was just publishing his famous manual of histology; Franz Leydig, who was studying the structure and function of the reproductive cells; and Rudolf Virchow, who would forge the link between cellular misfunctioning and disease.

It was the brilliant Virchow who initially captivated the young Haeckel. Virchow was, in those years, gathering the knowledge that was to make him the father of modern cellular pathology—a field whose name derives from the title of a book that he was to publish in 1858. His intellectual stature may be judged by the fact that *Cellular Pathology* was hailed almost one hundred years later as one of four greatest medical books since Hippocrates.[9] Virchow's lectures were exciting, vital, and new; in a few years' time, they would form the basis of his book. The students responded eagerly.

> This course is so unique of its kind that it is impossible for me to give you a complete picture of it yet [Haeckel wrote his parents enthusiastically]. . . . The course treats subjects . . . which have not yet even been printed, and which have only recently been discovered by Virchow himself. For this reason, the rush for it is quite extraordinary. The very large amphitheatrical lecture hall, containing over a hundred seats, is completely filled. . . . Everyone tries, as far as he can possibly do so, not to miss a single lecture, because he can hear things which can neither be heard nor read elsewhere. . . . Although difficult, Virchow's delivery is extraordinarily fine; I have never yet come across such crisp terseness, such compact strength, such close consistency, such sharp logic, and withal such remarkably clear descriptions and attractive animation in the delivery so intimately united as here.[10]

Haeckel was overwhelmed by the incisive mind of his young professor, by the "clear and biting sharpness [of] the absolutely matter-of-fact man [who exhibited] strong contempt and highly refined derision for all those who think differently from him."[11] He tried, briefly and futilely, to emulate Virchow, whose cold, unemotional, and profoundly analytical personality differed so utterly from his own romantic, warmhearted, and enthusiastic style. Haeckel worked until he had writer's cramp on notes taken in Virchow's lectures and spent hours in the laboratory learning to "see microscopically."

But Virchow was not the only professor to whom Haeckel was drawn. He spent the summer of 1854 partly in Berlin and partly in Heligoland, on the North Sea, studying microscopic seas creatures under the tutelage of the renowned Johannes Müller, the great physiologist who had taught Kölliker, Virchow, and many of the other leading young scientists. Müller was a different sort of biologist from most, a leader in the movement that emphasized studying organisms where and as they lived, not dead and preserved in the laboratory. With improvised nets of fine linen gauze dipped into the sea from a small boat, Müller and his students gathered magically exquisite creatures for study. This style of science appealed to the painter as well as the scientist in Haeckel.

I shall never forget [wrote Haeckel] the astonishment with which
I gazed for the first time on the swarm of transparent marine animals that Müller emptied out of his fine net into the glass vessel;
the beautiful medley of graceful medusae and iridescent
ctenophores, arrow-like sagittae and serpent-shaped tomopteris,
the masses of copepods and schizopods, and the marine larvae of
worms and echinoderms.[12]

Before succumbing to this aesthetic mode of science, Haeckel made
one more attempt at adopting a Virchowian rigor. For the summer of
1856, he was Virchow's "Royal Bavarian assistant at the Pathologico-
anatomical Institute at Wurtzburg."[13] By now, perhaps, Virchow had
recognized Haeckel's tendency toward emotionality and its twin, slip-
shod work habits, and felt they needed curbing or perhaps he was simply obscenely insensitive to the young man's feelings. In any case, the
first task Virchow assigned his new assistant was to perform a postmortem autopsy on a fellow classmate, a friend who had succumbed to
tuberculosis. It was a grotesque, appalling endeavor even for one hardened to the stench and rank unpleasantness of dissecting unpreserved
bodies in the summer heat. Haeckel lasted the summer out and won
praise from his mentor for the quality of his notes, but they parted company in the autumn; Haeckel stayed in Würzburg and Virchow moved
on to even greater eminence in Berlin, to become the professor and
head of a new pathological institute.

Haeckel finished medical school, making several trips to the
Mediterranean to capture specimens for his doctoral dissertation on
crayfish and some additional studies on marine crustaceans. Haeckel
made a brief and desultory attempt to open a medical practice to please
his father; with office hours from five to six o'clock in the morning, he
had only three patients in a year of practice. Haeckel then embarked
on research on the radiolaria, the group of wonderful single-celled sea
animals with intricate and beautifully sculptured "tests" or skeletons
through the holes of which gelatinous pseudopods protrude. When he
returned to Berlin from the Mediterranean in May of 1860, to prepare
his masterful work for publication, Haeckel heard for the first time of
a startling new book by the Englishman Darwin.

Haeckel was poised at a critical juncture in his career and intellectual development. He had accumulated an impressive body of work on
radiolaria—one that would earn him an international reputation and
secure him a position at the University of Jena, where he most desired
to be—but had not yet published it. He was full to the brim with data

and had not yet started to theorize. He no longer believed in Christianity, and so was not offended by the antireligious implications of Darwin's work as it was perceived at the time; Haeckel's God was Nature itself. He was also young and strong and given to buoyant enthusiasms.

He was no fool; he knew that Darwin's ideas were being received skeptically in Berlin. But when he read the book for himself—and Bronn's appended criticisms—he could not dismiss Darwin's ideas as so many others were doing.

> It profoundly moved me [he wrote to his biographer, Wilhelm Bölsche] at the first reading. But as *all* the Berlin magnates . . . were against it, I could make no headway in my defense of it. I did not breathe freely until I visited [the anatomist Karl] Gegenbaur at Jena (June, 1860); my long conversations with him finally confirmed my conviction of the truth of Darwinism or transformism.[14]

With that, Haeckel rose to Darwin's defense and became his noble knight errant in Germany. Both the "grandeur in this view of life"[15] proposed by Darwin and the humble, commonsense style of Darwin's examples and logic appealed to Haeckel. Here was a bold, sweeping new theory that was stunningly original, powerfully explanatory, and yet could be grasped by any good countryman who knew about crops and domestic animals. Neither a meddling Creator nor a mysterious vital force was needed to explain the world any longer. It suited Haeckel's temperament and beliefs exactly.

In 1861, Haeckel was offered the position of *privatdozent* (lecturer) at the University of Jena with his friend Gegenbaur, and happily accepted. Writing up his radiolarian monograph for publication in 1862, Haeckel saw over and over again how Darwin's ideas illuminated his material. It was Haeckel's first major publication, and he chose to declare unequivocally for Darwin's ideas in it. He presented his taxonomy, or classification, of the radiolaria, pointing out the "numerous transitional forms that most intimately connect the different groups and make it difficult to separate them in classification, to some extent." Mentioning Darwin's work by name, Haeckel observed how important the question of the fixity or mutability of species was to his theory. He then arranged the radiolaria into a genealogical tree—or phylogeny, as it would be called now, using a term that Haeckel coined—showing how neatly the evidence fit with the interpretation that all radiolaria are descended from a single, primitive form. "At the same time," he hedged, "this does not imply in the least that all the radiolaria *must* have descended from this primitive form; I merely show that, as a matter of fact, all these very varied forms *may* be derived from such a common fundamental type."[16]

In a long note, Haeckel made his views plain:

> I cannot refrain from expressing here the great admiration with which Darwin's able theory has inspired me. Especially as this epoch-making work has for the most part been unfavorably received by our German professors of science, and seems in some cases to have been entirely misunderstood. . . . I must express here my belief in the mutability of species and the real genealogical relation of all organisms. Although I hesitate to accept Darwin's views and hypotheses to the full and endorse the whole of his argument, I cannot but admire the earnest, scientific attempt made in his work to explain all the phenomena of organic nature on broad and consistent principles and to substitute an intelligible natural law for unintelligible miracles. . . . Undeniably important as are the principles of natural selection, the struggle for life, the relation of organisms to each other, the divergence of characters, and all the other principles employed by Darwin in support of his theory, it is, nevertheless, quite possible that there are just as many and important principles still quite unknown to us that have an equal or even greater influence on the phenomena of organic nature. . . . The chief defect of the Darwinian theory is that it throws no light on the origin of the primitive organism—probably a single cell—from which all others were descended.[17]

By the time these words were published, Haeckel had risen to the more secure and prestigious post of Extraordinary Professor of Zoology and Director of the Zoological Museum at Jena University. In 1863, at the scientific congress in Stettin, Germany, Haeckel took an even stronger and more public stand: he set about rectifying Darwin's "error" in omitting application of his theory to humans.

> As regards man himself [Haeckel said early in his speech], if we are consistent we must recognize his immediate ancestors in ape-like mammals; earlier still in kangaroo-like marsupials; beyond these, in the secondary period, in lizard-like reptiles; and final-

ly, at a yet earlier stage, the primary period, in lowly organized fishes.[18]

He then proposed the metaphor of the evolutionary tree, whose roots lie in the remote past:

> The thousands of green leaves on the tree that clothe the younger and fresher twigs, and differ in their height and breadth from the trunk, correspond to the living species of animals and plants; these are the more advanced, the further they are removed from the primeval stem. The withered and faded leaves, that we see on the older and dead twigs, represent the many extinct species that dwelt on the earth in earlier geological ages, and come closer to the primeval simple stem-form, the more remote they are from us.[19]

Here were two clear challenges to traditional thinking. First, Haeckel asserted that humans evolved from the apes; second, he elaborated a striking metaphor that presented the history and diversity of life as a single entity subject to a single set of natural laws. Not incidentally, Haeckel's arboreal metaphor was a progressivist one, with the main trunk leading straight to the most advanced, most perfect forms: humans. Haeckel was a charming man, a well-respected scientist, and a compelling public speaker, but even he could not convince all of his listeners.

Who was it Haeckel challenged? The personification of establishment science was by then none other than Haeckel's old professor, Rudolf Virchow. At the 1863 meeting at which they both spoke—Virchow directly following Haeckel—they showed the first signs of their growing estrangement. Their differences had surely been latent for many years, for two such men, with utterly different temperaments and views of life, could not for long agree. As Haeckel rose to prominence, he became Virchow's rival for power and national and international fame. While Haeckel never, perhaps, possessed an intellect as incisive as Virchow's, he was an original and creative theorizer, with all the advantages of immense persuasive powers in speech and written language.

It grew to a bitter struggle for dominance and control, fought on the pretext of evolutionary theory. The issues were political, given special urgency and vibrancy by the personal relationship between the men.

Virchow could permit no freedom of thought or action to his acolytes: he never forgave one student for marrying without his permission. Virchow expected Haeckel to continue to treat him as the Herr Doktor Professor, whose dictates were to be accepted without demur. Haeckel found the idea of remaining subservient to his onetime professor— who, after all, had urged him to think independently and untrammeled by emotion, hadn't he?—equally impossible.

Conflict was inevitable, *preordained*. Unlike the situation in England, evolution transplanted to Germany had little to do with religion. The battle here had scientist pitted against scientist, both laboring in the cause of political conviction.

Freedom in Science

CHAPTER 5

Virchow had started life very differently from Haeckel, and it was his life story, as much as anything, that led him to interpret evolutionary theory in political terms, opposing it at every turn.

For a man whose scientific renown would glow brightly more than a century after the work was completed, he came from a surprisingly ordinary family of Polish descent in Pomerania. Pomerania was at the time a rural province of Prussia still strongly influenced by the hereditary aristocratic class known as Junkers. Virchow's family was neither aristocratic nor wealthy; his father, a farmer and town treasurer, perpetually mismanaged money. Virchow secured a place at the Friedrich-Wilhelms Institute, where a few carefully chosen young men destined for careers as military physicians were offered free tuition. He squeezed every coin all the way through medical school, planning months in advance when he would be able to afford a new pair of trousers. He was acutely aware that scholarship students were looked down on as socially inferior. His letters home account for every expenditure, plead for additional funds, boast of his academic successes (which were abundant), and moralize in a rather coldhearted, opinionated fashion that seems odd directed from son to parents.

Intellectually gifted though he was—and most of his colleagues and peers quickly recognized Virchow's genius—he lacked the warmth, kindness, and skills for dealing with others that characterized Haeckel. Virchow was respected, even feared, for his brilliance and his intolerance of sloppy work or shallow thinking, but he was rarely loved.

Even his father complained of the emotional distance between them and received in response formal letters from Virchow admitting to his own "natural outer coldness and . . . withdrawn and self-limited disposition." As a birthday gift, he sent his father "a picture of our Royal couple entering the renaissance hall"[1]—an image that might be displayed in a shop or other public place as a pro forma demonstration of loyalty.

People simply never excited Virchow's attention. Indeed, only two subjects seemed to arouse his passion: learning and politics. In another letter to his father that shows their ongoing discord, he wrote:

> You consider me an egoist; this is possible. But you complain of my overestimation of myself; this cannot hold good in the same measure. True knowledge is to be aware of one's ignorance; and I am painfully aware of the vast gaps in my understanding. That is why I never stop and stand still in any field of knowledge; I like to learn, but I defend my opinions out of conviction. . . . A great idea attracts me beyond all bounds. . . . All my time is spent in hearing, learning and repeating what are often very dry facts, and for my real interests I can scarcely save an hour, except at the cost of my health. But I continue to work hard even on tedious and undesirable matters, for they could well turn out to be my only means of subsistence.[2]

He was proud, prickly, selfish, and as ruthlessly hard on himself as he was on others for their flaws. His angular, ascetic figure and precise manner revealed his character as clearly as Haeckel's bluff, loud-voiced handsomeness told his. A small, wiry man, Virchow had owlish eyes that seemed devoid of eyelashes. His piercing gaze fixed its subject through pince-nez set firmly on his beaky nose.

Virchow established his professional reputation by attacking his elders, seeking out and exposing their shortcomings or faulty reasoning in a sort of intellectual patricide. The pitiless clarity of youth stayed with him his whole life long. He zealously ridiculed inaccurate observations of such well-known researchers, demolishing their theories of disease and replacing them with his own meticulously reasoned, well-supported ideas. He trusted no one else's observations, complaining:

> One must work everything through for oneself from the beginning, and this is so difficult that one at times loses heart. . . . If the results were not before me, namely that I am now regarded as an authority on matters scientific by everyone . . . I would per-

haps really have given up by now. I, who have worked for so short
a time and who am infinitely ignorant about so many things, I an
authority? It is really ridiculous! How little must they know who
ask someone as ignorant as I am for answers![3]

If his passion for learning had overtones of conquering knowledge and
slaying intellectual dragons, his politics were even more combative. He
joined wholeheartedly in the growing revolution against the government. Sent in 1848 to tour Upper Silesia, a region devastated by famine
and a typhus epidemic, Virchow laid the blame squarely on the government's shoulders, holding them responsible for the ignorance and
lack of sanitation and proper food that fostered the disease. His report
was offensively outspoken. One bureaucrat scribbled in the margins:
"This so-called frankness appears to me as total subjection to foolish
political fantasies. I had taken Virchow to be more reasonable in political affairs. . . . What an unfortunate waste of talent!"[4]
   He joined a radical group of physicians calling for massive reforms
in public health and policy, mostly expressed in a journal he edited (and
largely wrote) called *Die medizinische Reform (Medical Reform)*; he
also launched a more strictly biomedical journal, called *Archiv für
pathologische Anatomie und Physiologie und für klinische Medizin*,
known after 1903 as *Virchow's Archives*.
   In 1848, he actually took up arms and fought at the barricades in the
streets as the revolution became, briefly, a military reality. When the
government regained control, Virchow was dismissed from his job at
the Charité Hospital. Reinstated "on probation," he did not regain his
free room and board. Over the next five hungry months, his sense of
martyrdom grew while the government sought ways to remove him from
Berlin. Eventually he was offered a position at the University of
Würzburg in Bavaria, on the condition that he would not make
Würzburg a "playground" for his radical tendencies. Cowed, at least on
the surface, Virchow gave his word, allowed his radical journal to lapse,
and adopted a more circumspect role he referred to as "observer."[5]
   The new, apolitical Virchow was a scientific terror. His incisive intelligence and meticulous observations soon earned him national and
then international renown. At the heart of his work lay two critical
ideas that predisposed him against evolutionary theory. The first, represented by the aphorism *Omnis cellula a cellula,* was that all cells
come from other cells. At the time, the spontaneous generation of living matter from inorganic remains was still a lively theory. No, Virchow
asserted, the body is made up of cells—the organism is, in fact, a "de-

mocratic cell state,"[6] a "federation"[7] of "equal individuals,"[8] each doing his or her own part—that undergo continuous, repetitive cycles.
Cells are not created anew, from nothing; cells beget cells which beget
cells, eternally. The system remains the same, self-generating, perpet-
ual, *except* when something goes wrong.

Here lay the second important tenet of Virchow's beliefs: pathology
was cellular misfunction, not an imbalance of the humors or a disorder
of the nervous system as others maintained. Although he had an intrinsic dislike of theories—"This is not the time of systems; this is the
time of research of details," he would write[9]—this creed was the cornerstone of his medical work. He tried to show that a cell that is different from its progenitors, a cell that does not fulfill its appointed role,
makes the entire cell-state ill. Every deviation from the parent type was
pathological and disease-causing; change itself was evil. In short, descent with modification—a term Virchow had not yet heard but which
approximated his phrase "life under changed conditions"—*was* the
very definition of pathology. Little wonder Virchow opposed Darwinism ferociously when it arrived in Germany!

But his ability to quash Darwinism was a direct function of the
change in fortune Virchow had undergone in the years immediately before Darwin's incendiary ideas reached German audiences. After his
exile from Berlin in 1849, Virchow built his reputation as a premier scientist steadily and methodically. He published papers full of new and
stunning observations, gave dazzling lectures, conquered conferences
until it became a positive embarrassment to the German government
that he was not in Berlin, the seat of all learning and power. His radical political beliefs were held in abeyance, breaking the grounds of discretion only occasionally.

In 1856, Virchow must have been smugly satisfied to receive the call
to return to Berlin as professor of the newly created chair of pathology. Now it was his turn to make the government squirm; he demanded
an entire institute be built for him and subsidiary academic positions
created that he would have the right to fill. His conditions were met and
he returned to Berlin and became an unstoppable force, soon known as
the pope or pasha of medicine. Indeed, he ruled his world like an oriental potentate—all-powerful, sometimes arbitrary, demanding the utmost of personal loyalty and sacrifice and rewarding lavishly those who
complied. He trained most of the next generation of researchers and
professors in various biomedical fields and personally engineered their
professional appointments. "When they speak of the German School
[of science], it is me that they mean," he observed regally.[10]

Virchow's appointment to Berlin marked the point of inflection in his climb to eminence in three fields: medicine, politics, and anthropology. In 1856, Virchow was already editor of two important medical journals and had begun publication of a six-volume work, *Handbuch der speziellen Pathologie und Therapie (Handbook of Special Pathology and Therapeutics)*. Two years later, publication of Virchow's greatest work, *Cellular Pathology*, consolidated his intellectual domination of German medical science.

Shortly thereafter, the political Virchow reemerged with his election to the city council, followed in 1860 by his appointment to the Wissenschaftliche Deputation (the Council of Scientific Advisers to the Prussian government) and in 1861 by his participation in the founding of the radical German Progressive party. He became a clever and outspoken opponent of the militant prime minister, Otto von Bismarck, fighting nearly all of his policies and empire-building schemes with bold speeches and ruthless criticism. Though Virchow was unsuccessful in foiling Bismarck, he proved a painful annoyance. Bismarck actually concocted an excuse to challenge Virchow to a duel, doubtless with the intent of maiming or killing him (as Bismarck, trained in military academies to fence and to shoot with great skill, had already done to other opponents). Virchow declined, it is said by invoking the proviso that he would duel only with scapels.

In anthropology, Virchow's fame and power grew apace. He was instrumental in founding the museums, scientific societies, and journals that pertained to a broad range of topics in archaeology and physical anthropology. Like many of the era, he was obsessed with questions of racial origins and migrations in prehistoric times, though he dealt with only relatively recent and anatomically modern remains. His deep opposition to the Darwinian concept of descent with modification was a formidable barrier to its acceptance.

As important as Virchow's attitudes and power were to the ultimate reception of Darwin's ideas, the debate on human evolution was triggered before *The Origin of Species* arrived—not by a theory but by a fact as concrete and stony as any could possibly be.

In 1856, some workmen quarrying limestone found some large, dense, strange bones buried in the mud inside a narrow cave near Bonn known as the Feldhofer grotto. The grotto was in turn located in a valley known as the Neander Thal. The fossils were turned over to the local schoolteacher, Johannes Karl Fuhlrott, an avid antiquarian and naturalist. He soon sought advice from Hermann Schaaffhausen, an anatomist at the University of Bonn.

The specimen was a fossil human, the first to be recognized as ancient in origin and nonmodern in anatomy. The skull was capacious, long and low, with eye sockets that glowered emptily beneath strongly projecting browridges; the thigh bones, or femora, were stout and strongly curved, but hauntingly familiar; and so it went with every bone in the partial skeleton. It was the tangible evidence of human evolution—the proof of the existence of another, older, more primitive sort of human.

Schaaffhausen and Fuhlrott presented the skeleton to the scientific community in early 1857. Initially referred to as the Feldhofer skeleton, this specimen was the original Neanderthal; when German spelling was rationalized at the turn of the century, to reconcile spelling with pronunciation, the colloquial name for this specimen (and those like it which would be found in the future) became, properly, Neandertal.[11]

Schaaffhausen delivered a longer, more dogmatic paper on the skeleton in June of 1857, to a highly skeptical audience. He described the specimen as unquestionably dissimilar to modern humans, indubitably older than the Celtic and Germanic races of Europe, and genuinely fossil: in short, a new and archaic type of human. Aside from the technical descriptions of its peculiar, elongated cranium with ominously protruding browridges, Schaaffhausen noted the massively built, stout, and strongly curved thigh bones and the evidence of a broken and poorly healed left elbow, which caused the bones of the left arm to be markedly smaller than the mighty right arm. In no other way was the fossil skeleton pathological, he asserted firmly (and he was correct in this).

Virchow and his followers replied swiftly to establish an opposing point of view. Given that pathology was deviation from the parent type—and Virchow explicitly applied this idea to the origin of races[12]—it was flatly impossible that *normal* Europeans were derived from a different type of human. In a delicious confusion of individual pathology with racial or species pathology, Virchow and his minions set about interpreting the Feldhofer Neandertal as a pathological individual rather than as a new species of human ancestor.

One of the critics was August Franz Mayer, who suggested that the curvature of the legs reflected both childhood rickets and a lifetime of horseback riding. The enlarged browridges, he postulated, reflected a habit of scowling or otherwise exercising the muscles of the forehead. As for antiquity, Mayer thought it far more likely that the skeleton was simply one of the Russian soldiers who traversed Germany in 1814 en route to attacking France. In Huxley's wry summary of Mayer's argument, the individual was simply

> a rickety, bow-legged frowning Cossack, who, having carefully divested himself of his arms, accoutrements, and clothes (no traces of which were found), crept into a cave to die, and has been covered up with loam two feet thick by the "rebound" of the muddy cataracts which (hypothetically) have rushed over the mouth of his cave.[13]

Ironically, this fossil provoked a rejection of evolutionary theory in Germany, yet would become proof of evolution in England.

While Virchow himself withheld published comment for some years, his tacit rejection of the Neandertal remains' normalcy and antiquity was well known. He later actually endorsed Mayer's interpretation of the bones as rachitic. This was an amazing lapse of intellectual honesty for Virchow. The leg bones are very thick and boast an exceptional quantity of bony tissue (as do all Neandertal bones) rather than showing the thinned, porous and calcium-depleted condition typical of rickets. He cannot have overlooked this fact, since Schaaffhausen made explicit mention of the bones' thickness and density, nor is there a chance that Virchow was unfamiliar with the disease. Rickets was a common nineteenth-century ailment about which he had written two important papers in 1853 and 1854.

The only explanation for his failure to observe and understand the visible evidence can be the power of his various beliefs. First was his conviction about the nature of pathology—life under changed conditions or the failure of cells to produce other cells just like themselves. Clearly these Neandertal bones were not like those of modern humans so, logically, either they were pathological or (unthinkably) modern humans were. This interpretation was aided and abetted by the presence of a genuine pathology—the once-fractured arm bone—that had nothing to do with the anatomy of the rest of the skeleton.

Second were the political implications of accepting the deceased Neandertal as a normal individual; it would imply that transformism was correct. Transformism was a rather Lamarckian view of the mutability of species that preceded Darwinian evolution in Germany, France, and elsewhere. What connected the two theories was the essential belief that life-forms had changed over time; what separated them was the proposed mechanism, which for Darwin was natural selection and for transformists was a vaguely described will or yearning of the organism for self-improvement. However, transformism was the scientific equivalent of the French Revolution: a dangerous doctrine of the possibility of change in social as well as biological spheres. Much as he yearned for political reform, Virchow was inevitably opposed to an idea that might spawn chaos.

When Darwinism came to Germany, only a few years after the Neandertal skeleton had been described and debated, those who supported evolutionary theory (like Haeckel) immediately realized it was much more than a scientific hypothesis, as Virchow did. But the response of Haeckel and his followers was positive, not negative. In the words of one historian of science, Darwinism was to Haeckel "a complete and final rendering of the nature of the cosmos . . . [a means of studying] the world and everything in it including man and society as part of an organized and consistent whole."[14] It explained human biology as well as human behavior and human society. As Haeckel began developing his ideas about government and social policy, based on Darwinism, he forged a strong link between the fossils, evolutionary theory, and political ideas. These rather dangerous extensions of Darwinism served as the catalyst, hardening Virchow's denial of the existence of fossil humans into a lifelong crusade.

If transformism's imprecise reasoning was repugnant to Virchow, how much worse was Darwin's diffuse style of illustrating his thesis by anecdote after anecdote, one rambling natural history story after another. That Darwin's concept of natural selection was a powerful explanatory model might even have made things worse, for the idea of descent with modification and life under changed conditions *as the norm*, as the healthy rule and not the pathological deviation, would have been unacceptable to Virchow. In Germany, as in England, Darwin was criticized for his carelessness about footnotes and proper attributions: another amateurish foible that the meticulous and precise Virchow would have detested.

Added to Virchow's personal distaste for Darwin's literary style was the fact that Haeckel's championing of Darwin's hypothesis amounted to a challenge to Virchow's domination of German science. As Haeckel elaborated his ideas along Darwinian lines, and propped up the intertwining vines of his own Romantic notions with the stakes of natural selection, Virchow grew determined to weed them all out. Denying the fossils became part and parcel of the strategy to root out the ideas.

The scientific congress at Stettin in 1863 marked a crucial turning point in the relations between Haeckel and Virchow. Here Haeckel's campaign for the acceptance of Darwinian evolution became overt, tying Darwin's ideas closely to *Naturphilosophie* and to Haeckel's own rather emotional and unscientific view of the world. Darwinism was the heart of a far-reaching philosophy that explained human social institutions as well as physical attributes. It was "only a small fragment of a far more comprehensive doctrine—a part of the universal theory of development which embraces in its vast range the whole domain of human knowledge."[15] This more comprehensive doctrine was a theory Haeckel later called Monism, a name that symbolized the oneness of all the universe. Monism stood in explicit contrast to dualism, the philosophy that held that the material and spiritual or intellectual worlds were separate but parallel. The sweep of Darwinian theory was bad enough, in Virchow's eyes, but the still broader brushstrokes of Haeckel's "advanced Darwinism"—what came to be called social Darwinism elsewhere—were intolerable.

There were other sources of offense to Virchow. Haeckel's tree of evolutionary descent was rooted in an original, single-celled organism. Naturally this organism was in turn derived from the inorganic, since everything in the universe—matter, soul, organic and inorganic alike—operated by one grand set of rules. Thus Haeckelian evolution incorporated the spontaneous generation of living matter from the nonliving as the solution to the "origin of life" problem. Finally, Haeckel's speech was strongly anti-Church and highly political; he proposed that Darwinism showed that change was an essential characteristic of history, necessary for progress, and that it was time to overthrow the "tyrants" and "priests."[16]

At that same conference, Virchow took a different tack, trying to divorce science from philosophy or religion entirely. In fact, his speech can be read as little more than a warning to his fellow scientists (Haeckel in particular) to stay out of the sphere of the philosophers and stick to the plain facts—as if data could be collected, interpreted, and presented devoid of theoretical context. Virchow apparently believed this stance was possible but he demonstrated otherwise in his own works. Virchow was certain that different methods applied to different fields: science required data and tests, whereas philosophy and religion were the appropriate realm for speculation devoid of the need (or possibility) of scientific proof.

For Virchow, then, this interesting concept of transformism or "this most fertile idea . . . an energetic ferment"[17] of Darwin's—utterly unproven and unsubstantiated—was entirely worthy of cool, unbiased investigation. It could be examined, discussed, inspected, and perhaps even tested, but it had never earned unthinking, wholehearted endorsement, such as he felt Haeckel was offering.

Virchow genuinely feared a return to the Romantic view of science that *Naturphilosophie* had engendered in his youth. Haeckel's subsequent work did nothing to dispel this fear, for he was also concerned with the evolution and inheritance of the soul. ("The *soul*??!!" one can almost hear Virchow snorting.) In a warped parallel to Virchow's cell-state, Haeckel wrote and spoke of the state-soul—the soul of the whole organism—hypothesizing that it is the amalgamation of the myriad individual cell-souls that creates the larger entity, the human mind or psychic individuality. In publications through the 1860s and 1870s, Haeckel suggested that each cell is composed in turn of smaller units, "plastidules," each of which possesses *its* own soul which is endowed with memory (the root of heredity) and the capacity to learn (the root of adaptation). Although it is possible, now, to cast this theory as a sort of precursor of modern genetic theory, the utter lack of evidence and the freedom with which Haeckel created new entities and then took them as established fact horrified Virchow.

Haeckel's boldness was also demonstrated by another, more famous "invention": the missing link, which he named *Pithecanthropus alalus* (speechless ape-man), whose fossil form would be found—Haeckel predicted—to combine elements of apes and humans. The charm of Haeckel's missing link is that it inspired a young Dutch anatomist, Eugene Dubois, to go to Java, seeking and *finding* exactly such a fossil—what we now recognize as *Homo erectus*. This stunning success cannot hide the fact that Haeckel's creature was created, described, and named without any evidence whatsoever as to its existence. Virchow the pragmatist, the data-based scientist, found this approach to science insupportable, especially as (to his mind) the existence of any form of fossil human different from living humans had yet to be established.

Unlike Haeckel, Virchow was wary of attacking powerful public institutions too openly. He had not yet forgotten the power of the state to take away the livelihood of young scientists who upset the status quo; he had learned circumspection at a difficult school. Agnostic though he was, Virchow chose not to oppose the power of the Church in 1863.

When in 1877 Virchow and Haeckel once again came into conflict on the podium at a scientific meeting, their enmity had reached its acme. Haeckel had grown ever more speculative in the intervening years, using his charisma and talents as a communicator to build a tremendous following throughout Germany and, indeed, most of Europe. His popular books touching on creation, evolution, the soul of mankind, and endless other topics sold in the hundreds of thousands.

In his 1877 speech, published as "Freedom in Science and Teaching," Haeckel urged a complete revision of the school curriculum, so that it would focus upon evolutionary theory as the unifying principle of both nature and society. His aim was to bring about the transformation of the Germans into a "unified people" with a common outlook and to rid themselves of the "outspoken sickly aversion . . . [the] human degeneracy"[18] that resulted from accepting the anachronistic, corrupting, and monstrous beliefs of traditional, Christian civilization. It was this perversion or contamination of the true German people and the true German ideals that led to the weakness and trouble of his day. Germans must unite, purify, and assume their rightful (superior) place in the world.

As always, it was not just education but social policy that was at issue. One of the aspects of traditional beliefs that was targeted by Haeckel for intense criticism was the notion of the equality of mankind. The stereotype of the superior Aryan or Nordic race—blond, strong, intelligent, moral, and brave—had been articulated by Arthur de Gobineau in a book, *The Inequality of the Human Races*, in 1853. As the colonial empires of the European powers expanded, the question of what to do with and how to treat indigenous peoples escalated from a theoretical discussion to a hotly contested issue whose edicts were strained daily by cross-cultural conflicts that were fueled by differences in expectations and mores. Haeckel identified Aryans with the true Germans, with the *Volk* of Germany imbued with power and intrinsic goodness and mystically tied to their holy German landscape. By the late 1870s, Haeckel held enormous, even unparalleled, sway over public opinion and German cultural attitudes.[19] He was perfectly positioned to effect a radical overhaul of the German educational system.

In contrast, Virchow still occupied the command center of German biomedical science, but he enjoyed no widespread public adulation and his ideas evoked no deep, emotional response from the people. Even his political activism was somewhat ineffectual, if brave and well-meant. Astonishingly, Virchow was now working with his long-time enemy Bismarck, in an unholy alliance against the Catholic Church. The trigger had been the announcement of the doctrine of papal infallibility (under prescribed conditions) in 1870, which had shocked Virchow deeply. This doctrine was rendered dangerous by the fact that the Church virtually controlled education in Germany. The issue became, in Virchow's mind, nothing less than a struggle for civilization, a *Kulturkampf,* to use the name he coined for a movement that was little

more than a thinly veiled effort to disembowel the Catholic Church in
Germany. Whereas Virchow aimed for a true separation of Church and
state (reminiscent of the separation of philosophy and science that he
had advocated in 1863), Bismarck's goal was to subjugate the Church
to his version of the state—a difference of intent that presumably escaped Virchow's notice at the time. Bismarck had already conquered
and subjugated vast territories that were once independent, fusing them
into a new German Reich, and he was beginning to use anti-Semitism
to manipulate the laboring masses into a pliable, reactionary group.
But the Catholic Church threatened Bismarck's intended unity, especially while the Pope could command complete obedience on any matter about which he chose to invoke the doctrine of infallibility.
Bismarck adjudged that the Pope's power must and could be broken
and Virchow, for once, agreed.

While this turn of events might have softened Virchow's antagonism
to Haeckel, who also openly opposed the Church, it was not so. Haeckel's increasing mysticism and evolutionary "religion" was little different from the Church's insistence on faith, not reason. Virchow turned
back to the inflexible insistence on matters of fact and data that had
formed the basis of his anti-Haeckel remarks in 1863. Symbolically,
Virchow was slashing desperately at a seemingly invincible Hydra with
heads labeled "Haeckel," "The Church," "mysticism," "Aryanism,"
"faith," and "Darwinism."

Virchow published his address of 1877 as "The Freedom of Science
in the Modern State," choosing a title that closely echoed Haeckel's
"Freedom in Science and Teaching" of the same year. The pointed difference was that Virchow was talking about the freedom of science *in
a political context*. In this work, Virchow flatly opposed the teaching
of Darwinism (or Haeckelism, as it was being called in Germany) in
the schools. He began with a statement of political strategy, a lesson in
expediency. The *Kulturkampf* had been only partially successful in
diminishing the Church's power. Rather than risk the chance to
strengthen natural sciences in the curriculum by pushing for this extreme, Virchow reasoned, it was better to exercise self-restraint and
work for the teaching of less controversial matters.

> We must draw a strict distinction between what we wish to *teach*
> and what we wish to *search for*. The objects of our research are
> expressed as problems (or hypotheses). . . . The investigation of
> such problems . . . cannot be restricted. . . . This is *Freedom of
> Enquiry*. But the problem (or hypothesis) is not, without further

debate, to be made a *doctrine*. In our teaching we must keep to that lesser but still large province, which we have really mastered.

Gentlemen, I am persuaded that only by such resignation, imposed by us on ourselves and practised towards the rest of the world, shall we be able to conduct the contest with our opponents and to carry it on to victory. Every attempt to transform our problems into doctrines, to introduce our hypotheses as the bases of instruction—especially the attempt simply to dispossess the Church, and to supplant its dogmas forthwith by a religion of evolution—be assured, gentlemen, every such attempt will make shipwreck, and its wreck will also bring with it the greatest perils for the whole position of science.

Therefore, gentlemen, let us moderate our zeal: let us patiently resign ourselves always to put forward, as problems only, even the most favourite problems that we set up; never ceasing to repeat a hundredfold a hundred times:—"Do not take this for established truth; be prepared to find that it is otherwise; only for the moment we are of the opinion that *it may possibly be so*."[20]

This argument appears rational, even moderate, but it conceals an underlying and passionate truth: Virchow did not and could not accept evolutionary theory. He did not for a moment believe in his heart that it could possibly be so, even after Huxley's masterful demonstration of the tremendous similarities among humans and apes as evidence for common descent. To Virchow evolutionary theory was virtually fact-free. He continued:

I am bound to declare that every positive advance which we have made in the province of pre-historic anthropology has actually removed us further from the proof of such a connection [between man and apes].

Anthropology is at present occupied with the question of fossil man. . . .

As recently as ten years ago, whenever a skull was found in a peat bog, or in pile dwellings, or in ancient caves, people fancied they saw in it a wonderful token of an inferior state, still quite undeveloped. They smelt out the very scent of the ape: only this has continually been more and more lost. The old troglodytes, pile-villagers, and bog-people, prove to be quite a respectable society. They have heads so large, that many a living person would be only too happy to possess such. . . . Nay, if we gather together the

whole sum of the fossil men hitherto known, and put them parallel with those of the present time, we can decidedly pronounce
that there are among living men a much greater number of individuals who show a relatively inferior type than there are among
the fossils known up to this time.[21]

In case his denial of the evidence is insufficiently damning, Virchow
restated his main point in capital letters:

> But one thing I must say—that not a single fossil skull of an ape
> or of an "ape-man" has yet been found that could really have be-
> longed to a human being. . . . WE CANNOT TEACH, WE CAN-
> NOT PRONOUNCE IT TO BE A CONQUEST OF SCIENCE,
> THAT MAN DESCENDS FROM THE APE OR FROM ANY
> OTHER ANIMAL.[22]

The fossils from the Neander Thal, supplemented by this time by another Neandertal jaw from La Naulette, in Belgium, and a complete
cranium from Gibraltar, failed to convince Virchow of the existence
of ancient and nonmodern humans (as would all of the additional fossil material that accumulated before his death early in the twentieth
century).

As he closed his argument, he took a final jab at Haeckel, whose popular works on evolutionary theory appealed to a wide audience who remained largely unaware of his sloppiness with the exact facts.

> [H]erein lies the great difficulty for every student of nature who
> addresses the world at large. Whoever speaks or writes for the
> public is bound, in my opinion, to examine with twofold exact-
> ness how much of that which he knows and says is objectively
> true. He is bound to take the greatest possible care that all the
> merely inductive generalizations which he makes, all his extend-
> ed conclusions according to the laws of analogy—however obvi-
> ous they may seem—be printed in smaller type under the text and
> that in the text itself he put nothing but what is really objective
> truth.[23]

Sacrificing Darwinism on the altar of political expediency was no hardship for Virchow; it was a welcome excuse to get rid of a troublesome
and utterly nonsensical theory. With one slashing blow, he tried to de-

capitate Haeckel's mysticism, the speculative pseudoscience of evolution, and the Church.

The truth is, for better or for worse, Virchow's science was always the handmaiden of politics and social reform. The obvious cases involve his work with sanitation, sewerage, and other public health measures, but anthropology and evolution were bent to his wider sensibilities, too. His 1877 speech was only one of several attacks on the ideas he detested.

Among those hated concepts was the Aryan myth, so in 1876 Virchow attacked it by starting a massive anthropological survey of the color of hair, eyes, and skin of 6,760,000 German schoolchildren. It was a bold attempt to transform physical anthropology from an eccentric and amateurish obsession with collecting peculiar skulls into a quantitative and modern science: a transformation similar to that which he had effected in human pathology. Perhaps Virchow also sensed the vicious turn racism—in particular, anti-Semitism—was starting to take under Bismarck's encouragement, with Haeckel's growing "scientific" justification. Perhaps, equally, Virchow was uncomfortable with his own obviously Slavonic ancestry, revealed in both his name and his appearance. Whatever the motivation, he set out to destroy the myth of the tall, blond, blue-eyed Aryan—a prototype for which his arch-enemy Haeckel might have modeled. Children were inspected in their schools, a procedure that gave rise to chilling rumors. One description of the survey recounts:

> The report had spread far and wide that all Catholic children with black hair and blue eyes were to be sent out of the country, some said to Russia, while others declared that it was the King of Prussia, who had been playing cards with the Sultan of Turkey and had staked and lost 40,000 fair-haired, blue-eyed children, and there were Moors travelling about in covered carts to collect them; and the schoolmasters were helping for they were to have five dollars for every child they handed over. For a time popular excitement was quite serious. . . . One schoolmaster who evidently knew his people assured the terrified parents that it was only the children with blue hair and green eyes that were wanted—an explanation which sent them home quite comforted.[24]

The populace obviously feared that officially condoned discrimination and exile, or even death, on the basis of looks or religion was pending.

The wide currency of such rumors in the late 1870s shows how deep the roots of the Nazi's Final Solution lay. Too, the fatuousness of the schoolmaster's reassurance shows the disdain of the educated for the uneducated, who might not understand the truth if they were offered it but who (it was apparently believed) would credit utter nonsense if it were delivered authoritatively.

Virchow's survey, eventually published in 1886, proved his point: Germans were neither predominantly fair of hair or skin nor blue-eyed, although Jewish children (surveyed separately) were shown as a group to be darker than Christian Germans. Virchow's research did not make even a small dent in the armor-plating of the Aryan ideal—it remained the concept of what a good German ought to look like—nor did it shake the fundamental conviction that Christian Germans were a racially unified group. He asserted, to no avail, "We know that every nationality, take for instance the German or Slavonic, is of a composite character and no one can say, on the spur of the moment, from what original stock either may have been developed."[25] That apparently casual choice of example—German or Slavonic—reveals his private concerns about his own heritage.

Virchow did everything in his power to control German physical anthropology and to block acceptance of Darwinism. As the years passed, Darwinism became more widely accepted and less controversial, especially in England, Belgium, and France. As more specimens were found, the fossil evidence for human antiquity, in the form of Neandertals, was also more generally admitted (except in Germany). In 1886, the discovery of two additional skeletons of Neandertals in Belgium, at a site known as Spy, confirmed—in most people's minds—the normalcy and antiquity of this type of fossil human.

But it was not so for Virchow or his followers. In 1889, he could happily give an address in which he stubbornly declared that evolutionary theory was all but defunct.

> Twenty years ago, Darwinism had just made its first triumphal march through the world. . . . At that time it was hoped that the theory of descent would conquer not in the form promulgated by Darwin but in that by his followers [i.e., Haeckel]—for we have to deal now not with Darwin but with Darwinians. No one doubted that the proof would be forthcoming, demonstrating that man descended from a monkey or at least from some kind of animal would soon be established. . . . This Darwinism has not succeeded in doing. In vain have its adherents sought for connecting links

which should connect man with the monkey; not a single one has been found. . . . It was supposed by faith, but it never belonged to science.[26]

Virchow continued to suppress and deny the fact of human evolution for his twin reasons: political expediency and his view of the proper conduct of science. But his power waned as he approached his death in 1902.

Haeckel continued to battle for evolutionary theory, defeating Virchow by longevity if not logic. Darwinism was for him, too, a political issue. At the turn of the century, Haeckel's influence grew more potent still with the publication of his most successful popular book, *Die Welträthsel (The Riddle of the Universe)* in 1899. Translated into twenty-five languages, *Die Welträthsel* sold more than 100,000 copies in the first year and, over the years, over half a million in Germany alone.[27] Daniel Gasman, a modern historian of science, explains the book's appeal:

In the *Welträthsel*, the scientific and philosophical "truths" of the world were seemingly presented in a clear and forthright manner by Germany's greatest biologist. The mysteries of the world and of life were, Haeckel assured his readers, readily explainable and within the grasp of science. . . . In addition, the *Welträthsel* offered a modern religious faith. The science of nature, Haeckel suggested, was not a vast impersonal discipline. On the contrary, it offered the basis for a faith as compelling as traditional religion.[28]

According to Gasman, Haeckel himself was regarded as godlike, a prophet, a genius who evoked a fervent nationalistic faith among his followers. His words were mystical, romantic, compelling: tantamount to holy. Virchow was dead and Haeckel's vision reigned.

In 1906, he founded the German Monist League devoted to bringing about his aims. Terrible doctrines, loosely based on Haeckel's interpretation of the meaning of evolutionary theory, succeeded each other, tumbling along in a frighteningly plausible stream of beliefs. Darwin's innocent theory was warped out of all recognition. At least Darwin, dead since 1882, was unaware of the ominous words being uttered in his name.

A distorted concept of human race was at the heart of Haeckel's theories. A race was, to him, not what biologists understand it to be today:

a regional subdivision of any species (human or not); a local population, loosely united by a tendency to share particular variations in phenotype (appearance) or genotype (genetic inheritance). To Haeckel, a race was a nationality, a tribe, or even an ethnic group who differed culturally but not genetically from their neighbors. Another race was "them" to "us," and all of Virchow's numbers showing that "we" were phenotypically diverse, too, had no effect.

Haeckel was completely open about his conviction that biological law must equally govern human society. "[N]atural selection in the struggle for life," he wrote, "transforms human society, just as it modifies animals and plants."[29] He flatly believed that races were as different from each other as species of animals, offering what seemed to be scientific support for outright racism.[30] He reasoned that, because the "lower races (such as the Veddahs [of India] or Australian Negroes) are psychologically nearer to the mammals (apes and dogs) than to civilized Europeans, we must, therefore, assign a totally different value to their lives."[31]

Because of the differences among races, maintaining (or reestablishing) German racial purity was of utmost importance to the well-being of the world. Germany, of course, represented the evolutionary apex of humanity and must lead the way in the "preservation of the individual character of nations and races,"[32] since the greatest progress would arise from the greatest peoples. In Haeckel's eyes, the nation was usually synonymous with the race and was also the sole effective unit in the process of social evolution.

Among other changes,[33] Haeckel's Monists proposed a graduated franchise, so that the votes of the more intelligent would count more than those of the ordinary laborer. Only by the subjugation of the individual to the good of the state could the proper order be established within the state and among states; only then could prosperity and happiness for the whole be achieved. Democracy was obviously doomed to disaster and so, too, was uncontrolled breeding by all elements of society without regard to fitness and worthiness. So strong were his beliefs that he was able to make extraordinary pronouncements without any apparent awareness of their patent flaws: the "destruction of abnormal new-born infants" as was done in ancient Sparta, was a "practice of advantage both to the infants destroyed and to the community," Haeckel said.[34] In what way destruction was an advantage to the dead infants is hard to imagine.

With the heat of his passionate "science," Haeckel welded together the race and the state.[35] The German race must be subjected to authoritarian power and enhanced by eugenics; he led thousands to adopt his point of view. He offered a new and potent scientific truth to ease the pains of a troubled nation in a difficult age, and the people listened. Haeckel was blind to the evils he was advocating and to the distortions of evolutionary theory in whose name he proclaimed.

Between them, Virchow and Haeckel defeated empirical science in Germany altogether. By using science as the weapon of political reform, the one was led to deny the physical evidence of evolution apparent to his eyes and the other to mutate, expand, and wrench Darwin's poor theory out of all recognition.

Part III
Evolutionary Racism

Survival of the Unfittest

CHAPTER 6

THE ENGLISH-SPEAKING WORLD was not immune to the intellectual and economic forces that made Haeckel's ideas so appealing in Germany. Since well before the publication of *The Origin of Species*, the philosopher Herbert Spencer had been developing concepts of society and government that were closely congruent with Darwin's evolutionary ideas as he eventually articulated them. Indeed, one phrase now almost wholly identified with evolutionary theory—survival of the fittest— was Spencer's, not Darwin's, invention. There is no question but that Darwin was heavily influenced by Spencer's thought.

Born in 1820, Herbert Spencer was the son of a schoolteacher and a sometime journalist in economics before an inheritance from an uncle freed him of the obligation to work for a living. He declined to go to university; in any case, he was a mediocre scholar in tasks that require memorization, such as languages. He did well in geometry, physics, and drawing. But he read widely and thought originally, perhaps because his lack of academic training left him unindoctrinated in the dominant paradigms of many of the fields in which he worked. In 1851, he published *Social Statics,* in which his ideas about the appropriate governing of society were sketched out. Other works followed, most notably his comprehensive work *The Synthetic Philosophy* (planned in 1860 and finished in 1896), the various volumes of which presented his views of biology, psychology, morality, and sociology. In it, Spencer elaborated his philosophical system, reworking Darwin's Spencer-influenced ideas back into his own constructs.

To Spencer, as to Haeckel, evolution was not simply Darwin's little theory about the origin of species; it was a universal principle. Spencer saw not a spindly vine, but a rampant, invasive Darwinism with curling tendrils that clung to every crack and bulge in the wall of life. Every aspect of reality was to be seen in terms of progressive development, he argued, from the evolution of the solar system to animal species and human society. Human products, such as art, science, industry, and language, also evolved, going from the lower, homogeneous stages to ever-increasing differentiation. Simplicity and uniformity give way to heterogeneity and "individuation." This was an inevitable and good process which functioned best when left largely undisturbed, Spencer believed, so he advocated an extreme form of laissez-faire government. Some might suffer along the way—indeed, it was inevitable that they *would* suffer—but the outcome was for the greater social good, the overall improvement of mankind and society.

It was a view of life wonderfully reminiscent of the Protestant perspective of previous generations. Worldly success had been accepted as prima facie evidence of goodness or divine approval; now that a scientific rather than a religious standard was to be met, worldly success was obviously an indicator of Darwinian fitness and genetic superiority. The poor and the working classes, who lived in squalor and misery, suffered because they were made of inferior stuff; their struggles were simply the manifestation of Nature's plan, of natural selection, and to interfere was to doom the society, the race, or even the species as a whole. Society and the economy would be led by those most fit to lead—those who had been selected over long generations of superior performance—and it was folly to contemplate any rapid or profound changes. The ideas meshed neatly with the long-standing practice of judging a new acquaintance by inquiring after his or her family; if you knew someone's "people," it was believed, you knew what tendencies and personality traits that individual was likely to have inherited.

Herbert Spencer was among the best-known of those who explored the implications of Darwinian evolution for other realms, though he was more popular in the United States than in his native Britain. Both countries were experiencing an industrial revolution that overhauled industrial production, urbanized the populace, and created ever-more-marked divisions among social and economic classes. The aristocracy, with its inherited lands and wealth, was soon joined by the nouveaux riches, the wealthy opportunists who had made the most of the technological revolution that was occurring. Together they formed a solid, conservative block deeply opposed to state interference in what were seen to be natural processes. This group argued that poverty laws, state-supported education, sanitary supervision, regulation of housing, tariffs, banking regulations, and even the postal system would impede the progressive improvement of society.

Of course Spencer was, like Darwin, fully acquainted with Malthus and the mathematical certainties of overpopulation followed by mass death. If the unfit, lower classes persisted in breeding, they would reap the inevitable consequences of their imprudence. He actually believed that the poor probably deserved to starve, but allowed that a minimum of charity so benefitted the character of the donors that it would probably do no lasting harm.

In short, Spencer was proposing that society evolved even as organisms did; Darwin's was a great truth that went far beyond the limited sphere for which Darwin had intended it. While inheritance in some unspecified fashion—the mechanism of genetics was still a mystery in the late nineteenth century—imparted variability to organisms within a species, Spencer suggested another source for the variability between individuals, classes, and societies or races. They, too, evolved toward perfection and they, too, were subjected to natural selection. According to his doctrine, there was a mysterious force, the Unknowable, that was continuously at work in the world producing variation, individuation, and specialization or elaboration. This Inscrutable Power, as he sometimes called it, was a shrewd concession to religion, for the Unknowable could be (and was often) readily identified with God. Spencer's was a sort of God-the-Watchmaker notion, a God who devised an amazingly intricate world, with built-in rules and movements of its own, who wound it up and allowed it to run without further intervention. The great Unknowable simply generated the initial diversity upon which natural selection would act mechanically.

In the 1870s and 1880s in America, similar social and economic changes nourished similar beliefs. The Civil War had left the population troubled, vulnerably aware of the differences among races, who were now declared legally equal, and the growing disparities among socioeconomic classes. The industrial North was on the rise, while the rural South struggled with reconstruction and a ruined economy. The robber barons of industry were flourishing, while others, formerly wealthy, faced destitution. Former slaves, most uneducated and impoverished, few with any idea of how to structure their own lives, found freedom a bitter fruit, not the milk and honey many had anticipated.

Among the educated classes, many turned to science for some sort of guidance and hope, specifically to the now widely accepted ideas of natural selection and Darwinism.

Spencer's words and work were well known in the States by the late nineteenth century, as witnessed, for example, by a popular lecture series on evolution sponsored by the Brooklyn Ethical Association in 1888. On the program were presentations on Spencer's and Darwin's ideas as well as a series of talks on such topics as the "Evolution of Mind," the "Evolution of Society," and the "Evolution of Morals." An impressive mix of self-proclaimed philosophers, scientists (including the famous paleontologist Edward Drinker Cope), and clergymen pontificated on that occasion.

But America was also growing its own philosophers of stature. Yale professor William Graham Sumner, a vigorous and influential man, sprinkled the fertilizer of the Protestant work ethic onto the soil of Malthusian economics in which Darwinian evolution was rooted. His words appealed to a wide audience. Though Sumner disliked the label "social Darwinist"—it was generally used as a derogatory term—his beliefs certainly resembled those of Spencer. It was readily apparent to him that people were biologically and, more important, morally unequal. Natural selection must be left to take its course, he maintained, and he felt a deep animosity for the social "meddlers" who were always trying to help the unfit. Poverty was simply the natural result of innate inferiority and a lack of thrift, industry, honesty, and sobriety. To alleviate it was to encourage bad behavior, to foster the spread of undesirable traits in the population, and to burden unfairly those who had succeeded through hard work. The only justifiable role for government was to defend the property of men and the honor of women. Natural selection might be harsh, Sumner believed, but it was just. He once quipped that the only alternative to survival of the fittest was survival of the *un*fittest. Competition was glorious, the new salvation that would improve the lot of all.

Although they would have defended the distinctness of their views, Spencer and Sumner had much in common. Both appealed to the conservative middle-to-upper classes; both opposed social programs and government regulation in economic areas; both, in fact, were using Darwin's theory as a scientific justification to prop up their social and economic policies. By anchoring their ideas in science—the new religion of the age—both Sumner and Spencer imparted a sense of inevitable rightness to their convictions. Progress was destined, and it was regrettable but only to be expected that it would occur at the expense of some and to the benefit of others. While Darwin himself had certainly derived some of his ideas about nature from the predominant Victorian rhetoric, his was a theory of biology, not politics. But now Darwin's concepts, based on the breeding of pigeons and the diversity of primroses, were recycled back into social theory with a vengeance. These migrations of Darwinian evolution out of the biological sphere and into a much broader one echoed the duel between Virchow and Haeckel slightly earlier in the century. The political implications of Darwinism had always been foremost in Germany.

In the intellectual and social climate that prevailed in England and the United States, it was a short step from social Darwinism to eugenics, a term coined in 1883 by Francis Galton, Darwin's cousin. Galton, a scientist and mathematician, relied on a Greek root meaning "well-born" to express his concept of a science that would improve the human species by affording "the more suitable races or strains of blood a better chance of prevailing speedily over the less suitable."[1] Even though genetics was deplorably undeveloped, Galton enjoyed a Victorian optimism that science and determination would conquer all. Like Darwin, Galton was thoroughly familiar with the impressive effects of careful breeding as a means of bettering crops and domestic animals alike. "Could not the race of men be similarly improved?" he asked innocently. "Could not the undesirables be got rid of and the desirables multiplied?"[2]

Galton was almost exactly the same age as Spencer and Huxley, some thirteen years younger than his cousin Darwin and a member of the generation of bright young men who found *The Origin* a revelation. "Your book drove away the constraint of my old superstition," he wrote Darwin, "as if it had been a nightmare."[3] By the time he read Darwin's words, however, he had been on an intellectual and geographical odyssey that left him open to these new ideas.

Galton was born to a Birmingham family whose wealth was based initially on gun manufacture and, in his father's generation, on banking. His mother was the daughter of Darwin's unconventional grandfather, Erasmus, a man who had proposed his own rather vague theory of evolution. Galton showed tremendous intellectual abilities from an early age—he could read at two and a half and wrote and did arithmetic at four—but his attempts at gaining a conventional education were unhappy. He seemed driven to excel, even as a youth, but at a cost to his health and well-being. Intending to follow in Erasmus's footsteps and become a physician, he enrolled at King's College Medical School in London, to be plagued by severe headaches that did not prevent him from compiling an excellent record. In 1840, he switched to Cambridge University to pursue a mathematics degree, a stressful endeavor that provoked a nervous breakdown in his senior year—the first of several during his life. He completed the degree, but failed to obtain honors, and resumed medical studies with no real zeal. By 1845, the pressure slackened: his father had died and Galton had come into a sufficiently large inheritance that he was never again obligated to work for a living. Galton left for Egypt with two friends.

This was the first stop on a series of meandering travels that were important in the formation of Galton's character. Rather than dawdle in the comfortable surroundings of Alexandria, or float down the Nile on a luxurious cruise, Galton crossed the desert to Khartoum, traveling by camel and Nile barge. He wandered on to Beirut and Jerusalem via an extended stay near Damascus, where he learned Arabic. A few years after returning to England, he undertook another prolonged journey: an exploration of southern Africa in the region where the Namaqua and Damara peoples were at war. He negotiated a Pax Britannica, impressing the Namaqua chief with his red hunting coat, and brought back to the Royal Geographical Society a wealth of important detail about the hitherto unmapped area, earning a gold medal and a membership in the Royal Society in the process.

Like many another Victorian explorer or traveler, Galton's time in the colonies seems to have been somewhat sybaritic; he refers to it himself as a "very oriental life"[4] in the Middle East. The family has apparently destroyed most of his letters from this trip, but records remain of his negotiating to buy a beautiful Abyssian slave girl who was probably a concubine. These trips provided him with a firsthand and certainly intimate look at life as it was lived by those of other races.

The influence of his experiences abroad was long-lasting. Despite his obvious fondness for many of the non-Europeans among whom he spent his time, Galton was not exempt from the racism typical of Englishmen of his class and era. The ambivalence of his feelings is neatly illustrated by his actions and their consequences. He was sufficiently intimate with someone on this trip—perhaps the beautiful Abyssian slave girl—to catch a venereal disease during this interlude. Though he received medical treatment, the disease apparently rendered him sterile. No record remains of that "wife" or her eventual fate. Galton's later, extremely proper and public marriage to Louisa Butler, the daughter of the headmaster of Harrow and Dean of Peterborough Cathedral, was childless—a curious and ironic circumstance for an intellectual whose main topic was eugenics.

As Daniel Kevles, a modern historian of science, has observed, Galton's fondness for sparsely settled regions applied to the scientific endeavors he undertook as well as to his travels:

> Although Galton resembled the typical scientific amateur of the nineteenth century in that he was untrained in the research he eventually pursued, he was atypically drawn throughout his scientific career to largely unpopulated fields, which in his day included both statistics and human heredity. If at times he embarked on a subject to which others had contributed, he did not begin his research by analyzing the existing body of scientific literature; his library contained hardly two dozen volumes acquired to forward his various inquiries. He learned from the work of others but did not approach it systematically. He came upon useful treatises by chance or sought them out as he happened to need them. . . . He was a rough-cut genius, a pioneer who moved from one new field to the next, applying methods developed in one to problems in another, often without rigor yet usually with striking effectiveness. Galton's innovativeness in science was intimately bound to his relative intellectual solitude—a propensity that arose from a measure of doubt in his abilities combined with a compulsion to excel.[5]

Whether living among other races abroad or attending the horse races at Epsom, one of Galton's strengths was his obsession with numbers. "Whenever you can, count"[6] was his motto. He scored women he passed on the street for their attractiveness, compiling a beauty map of England, and counted the fidgets in the audience at Royal Geographical Society meetings. Unable to resist the challenge of such extraordinary raw material in southern Africa, he measured the hefty bottoms of steatopygous Hottentot women indirectly, by persuading them to parade while he viewed them through his sextant. "I have seen figures that would drive the females of our land desperate," he wrote his elder brother, Darwin, "figures that could afford to scoff at Crinoline."[7] He enumerated everything that could be counted because he was enamored of numbers themselves. They held beauty, power, and truth in their patterns and arrangements.

The turning point in his career came with his book *Hereditary Genius*, published in 1869. It was an expansion of ideas he had expressed some years earlier in a popular magazine, in which he had raised the question whether natural ability—"those qualifications of intellect and disposition which . . . lead to reputation . . . [those attributes of] a leader of opinion . . . an originator"[8]—was inherited. He took a typically Galtonian approach to the issue. He derived a sample of eminent achievers in a two-hundred-year period from a wide range of fields—from poets to scientists, from musicians to military commanders—from biographical encyclopedias like the *Dictionary of Men of the Time*. Then he examined their relatedness, finding that certain families were disproportionately likely to produce such men. To his mind, this established beyond any reasonable doubt the fact that character, intelligence, and innovativeness were as hereditary as height, shape of nose, or prominence of ears. The logical conclusion was that it would be "quite practicable to produce a highly gifted race of men by judicious marriages during several consecutive generations."[9]

Denying vigorously any imputation that social prominence or wealth fostered accomplishment, Galton was convinced that the differences in human success simply reflected the quality of the breeding material. He proposed that the British government ought to foster felicitous unions, by examining the populace for hereditary merit and sponsoring public, celebratory marriages in Westminster Abbey for the genetically elite. Eugenics, whether purely voluntary or state-regulated, was the obvious, scientific, and appropriate means of improving the nation. Galton was the prophet of this new religion.

The time was ripe for such beliefs. In 1877, an American, Richard Dugdale, published his study of the Jukes family, which made their name synonymous with social pathology. They were a family riddled with prostitutes, criminals, paupers, and ne'er-do-wells that Dugdale had traced through seven generations, back to one singularly dysgenic couple in upstate New York. The study seemed to confirm everyone's worst suspicions about undesirables; few stopped to consider that environment, not heredity, might be at work (even though Dugdale himself allowed a considerable role to environment). In a pamphlet published later by the American Eugenics Society, the effects of allowing unchecked breeding of such a pair were made clear in terms of hard cash: the price put on segregating the original Jukes couple from the rest of society for life was "about $25,000"—a vast sum of money in the early 1900s. The pamphlet continued in a catechismal vein:

Q. Is that a real saving?
A. Yes. It has been estimated that the State of New York, up to 1916, spent over $2,000,000 on the descendents of these people.

Q. How much would it have cost to sterilize the original Jukes pair?

A. Less than $150.[10]

It was a chilling economic lesson and one that many Americans took to heart.

But in England, Galton badly needed actual data on the heritable characteristics of the population, not a single case history, however stunning its impact. In 1884, Galton set up an Anthropometric Laboratory at the International Health Exhibition at what would become the British Museum (Natural History) in South Kensington. Standardized measurements of height, weight, arm span, and so on were collected on over 9,000 people, including many sets of parents and children. He also published and distributed a questionnaire on heredity entitled *Record of Family Faculties*, with prizes for those who submitted information on large families of two generations. Working with these data, Galton sought, and found, a confirmation of the existence of what is known in statistics as the normal, Gaussian distribution: the familiar bell-shaped curve. A Gaussian bell curve is a graphic representation of the distribution within a population of any type of metric or measurable trait, such as foot length or head width. A few will be small, a few will be large, and the vast majority will cluster in the middle, making the hump in the curve. The most important feature of the bell curve is that the distribution of measurements falls symmetrically on either side of the mean or average value. The symmetry and predictability of the shape of the curve, across all sorts of traits, makes it possible to describe the patterns of deviation from the mean statistically, using a measure known as a standard deviation. By definition, 83 percent of all individuals measured will fall within one standard deviation of the mean. The two tails of the curve represent individuals who are more and more extreme—for example, who have very long and very short feet. And Galton believed that the distribution of genius or talent among humans followed a Gaussian curve just as the weight of sweet-pea seeds (the data on which he had actually tested this principle) did.

The anthropometric information he had collected was a perfect data set upon which to analyze heredity in this new, statistical way. He could estimate, for example, how many men of musical genius would be born in the year 1860 using the Gaussian distribution, since the shape and height of the bell curve remained constant from generation to generation. Playing with his data, Galton developed several important statistical techniques and principles, some having to do with measuring the extent to which one feature (such as head width) varies with another (such as head length). He did not, however, improve the acuity with which he distinguished among traits that were purely hereditary, like eye color, and those that were strongly influenced by socialization, environment, and training, like intelligence or morality.

Despite flaws in Galton's analyses and conclusions, eugenics quickly became a major concern in the English-speaking world. His thoughts on the urgency of establishing a scientific basis for human breeding epitomized the hopes and fears of the turn-of-the-century generation; his new statistical procedures lent authority to his pronouncements. That statistical concepts were difficult for the average person to understand was no deterrent and may even have enhanced Galton's reputation as a brilliant, learned man. Far from being tainted or discredited as racist or classist, Galton's eugenics was an ultimately respectable concern. Indeed, he was knighted in 1909.

Eugenics was rooted in the nineteenth-century confusion of national identities with races and nourished by the unease provoked by the bewildering array of social and economic changes that occurred at the turn of the century. Rapid industrialization gave rise to teeming cities; poor, ill-educated migrants and immigrants threatened to swamp respectable society; the calm, ordered, comfortable rule that had for generations separated class from class threatened to come apart at the seams. As World War I loomed nearer, morality, civilization, and security seemed ever more precariously perched on a trembling branch.

Though such concerns led to the ready acceptance of eugenic ideas, it was Galton who first brought eugenics into focus. But it fell to a younger man whom Galton befriended late in life, Karl Pearson, to push the movement further. Pearson was by far the greater statistician and had abilities Galton lacked for promoting eugenics.

Pearson was the son of an intensely ambitious Quaker solicitor, a cold and dominant man with little love or concern for his family. Karl was expected to follow his father into the law, but instead attended King's College, Cambridge, on a mathematics scholarship and went to Berlin and Heidelberg for postgraduate work. He started to prepare for the bar but in 1884, desperate to escape his father's steely plans for his life, managed to obtain a mathematics post at University College, London.

Pearson was already a social radical of sorts, believing that the mass of citizens must be subordinated to the good of the state, that morality was determined by the greatest good for the greatest number. He admired Spencer's notion of a laissez-faire government and regarded it as a natural outgrowth of Darwin's ideas, but to Pearson the most important struggle for survival occurred between nations or races rather than between individuals. The fittest nation would survive and, to this end, the British people must be improved. For Pearson, as for many others, *nation* and *race* were virtually synonymous terms; they referred to social and cultural groups, not biological entities as in modern thought.

Pearson's ideas were developed through conversation and lectures in his London circle of friends and acquaintances, which included sexual radicals George Bernard Shaw and Havelock Ellis, Sidney and Beatrice Webb, and Karl Marx's daughter. In 1885, Pearson founded the Men and Women's Club for open discussion of relations between men and women—and, as it turned out, for dabbling in free love and sexual experimentation. But there he also met Maria Sharpe, an intelligent, independent, and radical thinker with whom he fell in love. For her part, she found the notion of either conventional marriage or sexual relations so disturbing that she suffered a nervous breakdown following Pearson's proposal. (She recovered and married Pearson in 1890.)

Human heredity was still a problematic issue. Mendel's laws were rediscovered at the turn of the century, but exactly how inheritance operated was unclear. In the 1880s and 1890s, August Weismann proposed that there was a germ plasm whose continuity from generation to generation carried the hereditary material. He even identified this kernel of reproductive information with the chromosomes, which could be seen in stained cells examined through a microscope. Galton had arrived at a similar, if vaguer concept, of the "stirp" or root element that transmitted characteristics from parent to offspring. No acquired traits could be inherited if Weismann or Galton were correct, since these would not affect the germ plasm. But how did evolution *occur* if the germ plasm was passed on, unchanged, from generation to generation? The answer had to lie in the laws of probability and in the fertility of different segments of the species, as Galton had suspected but was too mathematically inept to prove.

These advances in knowledge of the hereditary mechanism did not bring a modern appreciation of the complexity of the system. We know now that hereditary traits may be affected by the interactions of many genes and, equally, by nutritional or environmental factors. For example, skin color in humans is genetically encoded, but predicting any individual's adult skin color is difficult. A child inherits more than a single set of genes for skin pigmentation—it is not just a matter of "swarthy" from the father and "peaches and cream" from the mother, for example—and genes for skin color seem not to work on a simple dominance scheme: one "dark" gene does not mask one "fair" gene. Regardless of the child's genotype, his or her skin color will also be influenced by nutrition, possibly by bouts of disease, and certainly by habits and occupation.

We also have a different perspective on what traits are heritable today. Many of the characteristics and behaviors that readily identify an individual as (most probably) coming from a specific region of the world may not be inherited at all, or may at best have only a small genetic component. Are Italians voluble and Finns taciturn? Sometimes. Is it because Italians more commonly carry genes for talkativeness than Finns? It is wildly improbable that this is so, for how could such a gene work? Any differences (if such exist) might better be attributed to the habits of their individual families and ethnic groups rather than to any national or racial genotype.

Pearson had a simpler view of what might be heritable and was more concerned with working out the statistics of heritability than defining its purview. He began collaborating with a fellow Cambridge graduate, Walter Weldon, who had embarked on a statistical study of shrimp and crabs. They found if the expression of particular traits was associated with a higher or lower death rate, then the fitness of an individual was roughly equivalent to the strength and direction of its deviation from the population mean for that trait. They worked together for years on this and many other subjects involving heredity and selection, leading to more than one hundred of Pearson's publications. The implications for human heredity were obvious and, in time, Pearson attacked this problem directly, reworking some of Galton's theories about the laws of inheritance.

Galton had found a disconcerting tendency that he called regression toward the mean. He noticed that the traits of offspring tended to fall closer to the average values for the population as a whole than to their parents' values—and this disposition posed a problem for understanding evolutionary change. If there were some inherent tendency to revert to the ancestral type, how exactly did evolution occur? Now Pearson showed that the only way to shift the entire curve in one direction or another—to improve, for example, the average intelligence of Britons—would be to increase the number of offspring born to those at the desirable extreme of the curve at the expense of those at the other extreme: eugenic breeding. It was simply a matter of allowing enough generations to pass for the results to become apparent. Galton exulted; Pearson turned to another of Galton's pet problems, the thorny issue of the heritability of intelligence.

Since no objective or even purportedly objective measures of intelligence existed, Pearson was forced to rely upon teacher's ratings of the temperaments and abilities of nearly four thousand schoolchildren. He found that the ratings given to siblings correlated admirably and about as closely as did physical characters: eye and hair color. Of course, if temperament and intelligence are heavily influenced by upbringing and environment, as we now believe them to be, siblings are as likely to share similar experiences as they are to share similar genes. Difficulties in comparing variables scored on a discontinuous scale (like eye color, which is blue, green or brown) with continuous variables (such as degree of intelligence) were swept aside. First Galton and now Pearson had come to the same conclusion, that intelligence and character were as heritable as eye or hair color or stature. Then Pearson conducted seemingly rigorous studies of the differential birthrates of various races or groups (including the feebleminded and criminal "classes") that did much to convince the masses that eugenics was important and accurate. The undesirables seemed to be breeding faster than the elite. Now it was time for the scientific control over human reproduction—the embodiment, they felt, of Darwinian ideals—to come into practice.

The resurrection of Mendel's work lent additional scientific credibility to eugenics. Soon institutions and organizations seemed to spring up all over, even though only a limited number of professionals were at work on eugenics topics. In 1904, Galton offered 500 pounds a year to University College, London, to fund a research fellowship in eugenics. Its first and only recipient, a young Oxford man named Edgar Schuster, founded the Eugenics Record Office on Gower Street in London. After a few years, it reverted to Pearson's leadership and—with Galton's sizable bequest of 45,000 pounds—this organization was transformed into the Galton Laboratory for National Eugenics. Galton's will also led to a Galton Eugenics Professorship (held, of course, by Pearson), a new Department of Applied Statistics, and the Galton Eugenic and Biometric Laboratories.

Data were collected on the frequency and distribution within families of an amazing variety of traits. One source lists "scientific, commercial and legal ability . . . hermaphroditism, hemophilia, cleft palate, harelip, tuberculosis, diabetes, deaf-mutism, polydactyly (more than five fingers), brachydactyly (stub fingers), insanity, and mental deficiency . . . alcoholism . . . [and] defective sight"[11] as traits on which a small army of volunteers provided information. These were by no stretch of the imagination all hereditary defects; we would now assign them to a bewildering mixture of causes, including genetic inheritance, environment (both in the sense of training and in the sense of contaminants or pollutants to which pregnant women may be exposed), infection, and behavior. Pearson employed a large staff—many of whom were women, as they could be paid less than men—to develop statistical techniques and carry out analyses on this overabundance of raw data, much of which was published in *The Treasury of Human Inheritance*.

But by far the most influential piece of research to which Pearson could lay claim was his study of differential reproductive rates, showing that the upper classes—the *sine qua non* of fitness—had fewer offspring than the lowest classes, the dregs of society. The good, the noble, the intelligent, the sensitive would, indeed, be conquered by the low, the selfish, the stupid, and the brutish, for the latter had sheer numbers on their side. The notion became such a truism that Theodore Roosevelt once chastised the middle and upper classes for committing "race suicide"[12] by limiting the size of their families. The science of eugenics, backed up by Darwinism and Mendelian genetics and lent urgency by such studies, began to enjoy considerable vogue.

Darwin himself was not so sure, at least when he first heard about eugenics. Upon reading Galton's book *Hereditary Genius*, Darwin wrote to his cousin: "You have made a convert of an opponent in one sense, for I have always maintained that, excepting fools, men did not differ much in intellect, only in zeal and hard work; and I still think [this] is an eminently important difference."[13]

As for the Eugenics Record Office, it was very similar to the eugenic registry that Galton had first proposed back in 1873; it might eventually have served as a repository of information about the fitness or worthiness of various individuals for breeding purposes that could be consulted when the prospect of engagements arose. Darwin's response to the idea was typically hesitant and ambivalent. The great practical difficulty, Darwin thought, "would be in deciding who deserved to be on the register. . . . Though I see so much difficulty, the object is a grand one . . . yet I fear [it is] utopian."[14] But his letters and papers show that he believed more heartily in eugenics as time went on—or perhaps he always had, but simply did not always care to acknowledge his thoughts.[15]

After Darwin's death in 1882, Francis Galton was one of the prominent men who choreographed the movement culminating in Darwin's burial in Westminster Abbey. It was a deliciously paradoxical occasion: a public celebration of the life and work of the man who had done more than any other to destroy the power of the Church, all to take place in the premier religious setting of England. The maneuvering entailed by Darwin's burial was conscious and deliberate; Galton was acutely aware of the propaganda value of such a public canonization of evolutionary theory, and who better to personify that theory than Darwin? Too, the symbolic reconciliation of the religious and scientific was both timely and important. Once berated as an enemy of morality and religion, Darwin was now sanctified and transformed into an icon acceptable to all aspects of society. His morality, patience, virtue, and greatness were praised by daily newspapers and religious publications alike. By implication, his once-revolutionary ideas were also absorbed into the interstices of the bedrock of establishment thinking. It had taken only twenty-three years.

Given the immense appeal of the eugenics movement to the conservative, well-to-do, and educated classes, it was no accident that Darwin's son, Major Leonard Darwin (retired) of Her Majesty's Royal Engineers, was also a devotee of the new vogue. It was, almost literally, in his genes. From 1891 to 1928, Leonard served as president of the Eugenics Society in England, during which time interest among the general public increased markedly. His participation provided the Darwinian endorsement that Galton, Pearson, and others realized would be so critical. Indeed, the dedication to Leonard Darwin's 1926 book, *The Need for Eugenic Reform*, reads:

> *Dedicated to the memory of*
> *MY FATHER*
> *for if I had not believed that he*
> *would have wished me to give such*
> *help as I could towards making his*
> *life's work of service to mankind,*
> *I should never have been led to*
> *write this book.*[16]

The intent to stamp the imprimatur of Darwin's by then legendary authority on the eugenics movement could not have been more plainly stated.

Sweeping Toward
a Racial Abyss

CHAPTER 7

By THE EARLY YEARS of the twentieth century, the main point of Darwin's theory had been completely transformed from origin of species into the principles of eugenics. Its central tenet was that humans could and should ensure the survival of the fittest humans, for the good of society. Eugenics, grafted onto the rootstock of Darwinism, blossomed and flourished in England. It, in turn, was transplanted happily to foreign soils, notably Germany and the United States. In these new environments, evolutionary eugenics—helped by the growing but still vague knowledge of the mechanisms of heredity—germinated and survived, drawing strength and character from the particular properties of these different environments. In short, the theory was evolving once again.

Across the Atlantic, it was Charles Davenport who brought eugenics to the people of America. A Harvard-trained biologist with an unusually strong—for the time—background in mathematics, Davenport was an archetypal East Coast Anglo-Saxon Protestant. He was born to a large and well-to-do family that was dominated by a rigidly authoritarian father. He rejected his father's ambitions for him to become a surveyor but absorbed his father's strict moral code and strong work ethic. As a young man, Davenport married Gertrude Crotty, a graduate student in zoology at the Society for Collegiate Instruction of Women (now Radcliffe College), who helped in his work and lost no opportunity to further her husband's career in biometry and eugenics. In 1899, he left his position at Harvard for one at the University of Chicago, where his reputation as a bright young man continued to expand. And, a few years later, he boldly approached the Carnegie Institute of Washington—a new research organization with a ten-million-dollar endowment from Andrew Carnegie—to set up an experimental station to study human heredity. His wish was granted and he became the director of a research station at Cold Spring Harbor, New York. His annual budget was lavish: $21,000, or better than twice Pearson's in London.[1]

Davenport's timing was excellent, for eugenics and the inheritance of various traits were subjects much on the mind of the public, especially the well-to-do like Carnegie. In the early 1900s, the United States experienced a large—and to many, alarming—rise in immigration, much of it from southeastern Europe. Absorbing this influx of people of different ethnic backgrounds into American society was sometimes tumultuous and difficult. In 1908, a playwright, Israel Zangwill, wrote of this process, coining a catchphrase and an image that were to reverberate for years to come: "America is God's crucible," Zangwill's character declared, "the great melting pot where all the races of Europe are melting and re-forming."[2]

Zangwill was a Jew, a Zionist, and an Englishmen, well known as one of the first writers to portray Jewish immigrant life in popular literature. But in much of the Western world, Jews were the target of substantial and freely expressed prejudice and distrust. To hear Zangwill celebrate the new American race that was being created in the melting pot did nothing to reassure the eugenicists. They were largely white, Anglo-Saxon, and Protestant, drawn from the upper and middle classes, and hoped desperately to protect their culture and privilege from these immigrant outsiders. In some minds, the melting pot was not a symbol of hope but was actually synonymous with the decline of the "white" race. This was precisely the theme of a popular book, *The Passing of the Great Race,* first published by an American, Madison Grant, in 1916.

Excerpts from this widely read work (which was translated and enjoyed healthy sales in Germany as well) show the astonishing extent to which prejudice, racism, and muddled science were accepted and acceptable. For example, Grant's views of the new immigrants and their impact on society are pithy:

What the Melting Pot actually does in practice can be seen in Mexico, where the absorption of the blood of the original Spanish conquerors by the native Indian population has produced the racial mixture which we call Mexican and which is now engaged in demonstrating its incapacity for self-government.[3]

The history of immigration into the United States was little more than a tale of strong beginnings followed by disaster, in Grant's eyes. "Native Americans" is a phrase Grant uses often, but not to indicate the American Indians who were displaced or killed by the wave of European colonists. Instead, in a usage common at the time but jarring now, Grant's "native Americans" are those western European colonists themselves as distinct from the more recent, post–Civil War immigrants. The group of indigenous peoples now commonly designated as Native Americans was of little or no concern to Grant.

> These new immigrants [Grant lamented] were no longer exclusively members of the Nordic race as were the earlier ones. . . . The transportation lines advertised America as a land flowing with milk and honey and the European governments took the opportunity to unload upon careless, wealthy and hospitable America the sweepings of their jails and asylums. . . . Our jails, insane asylums and almshouses are filled with this human flotsam and the whole tone of American life, social, moral and political has been lowered and vulgarized by them.[4]

Convinced that "moral, intellectual and spiritual attributes are as persistent as physical characters and are transmitted substantially unchanged from generation to generation,"[5] Grant left his readers in no doubt as to the eventual outcome of Darwinian selection if the "native American" stock is allowed to be so polluted. His thundering conclusion is strategically placed before almost 150 pages of appendices, maps, and documentary notes that are the invincible armor protecting his arguments:

> We Americans must realize that the altruistic ideals which have controlled our social development during the past century and the maudlin sentimentalism that has made America "an asylum for the oppressed," are sweeping the nation toward a racial abyss. If the Melting Pot is allowed to boil without control and we continue to follow our national motto and deliberately blind ourselves to all "distinctions of race, creed or color,"the type of native American of Colonial descent will become as extinct as the Athenian or the Age of Pericles, and the Viking of the days of Rollo.[6]

It was a racist, reactionary, inflammatory, and wrongheaded book that appealed to the worst fears of his readers. Sadly, it was also very popular and was reprinted and/or revised at least seven times before the start of World War II.

Davenport's view of the whole melting pot issue was similar but more subtle. He shared with Grant the commonplace view that race was the same thing as ethnic or national identity; like Grant, he also believed, unswervingly, that both physical and character traits were heritable. And Davenport's self-selected task in life was to protect and preserve the "national protoplasm" from threats of degradation and deterioration: exactly the intent of Grant's work. (In fact, Grant cited Davenport as the authority behind some of his seemingly dogmatic assertions.) But Davenport was liberal enough in his beliefs to recognize that it was wrong to write off entire races as undesirable or unfit: "No race *per se*, whether Slovak, Ruthenian, Turk or Chinese, is dangerous and none undesirable," he said.[7]

Davenport urged that immigration policy ought to be based on the hereditary history of the specific individual and his family, not on the place of origin. This attitude did not stop him from stereotyping Italians as prone to "crimes of personal violence," Poles as "independent and self-reliant though clannish," Serbians and Greeks as "slovenly," or Swedes and Germans as "tidy."[8] And, of course, he believed these behavioral traits were genetically controlled. He argued that no blending or melding of the traits of different races was possible in any case; all that could be done was to increase or decrease the proportion of the population that held certain genes. "The idea of a 'melting pot' belongs to the pre-Mendelian age. Now we recognize that characters are inherited as units and do not readily break up."[9]

The next obvious step was to begin massive studies to determine exactly which characters were heritable, so useful decisions on the eugenic fitness of individuals could be made. Davenport advocated not only positive eugenics (encouraging the fittest to have more children) but also negative eugenics (actively discouraging the breeding of the less fit). Davenport believed that it was both justifiable and sound to sterilize those carrying "defective protoplasm," when they could be identified. In general, however, he thought sexual segregation of the unfit was a less expensive course.

With a small but select staff, Davenport enlarged his own studies of skin, hair, and eye color, as well as attacking a raft of more serious issues: brachydactyly, polydactyly, albinism, hemophilia, otosclerosis, color blindness, Huntington's chorea, insanity, epilepsy, alcoholism, pauperism, criminality, tuberculosis, goiter, and feeblemindedness, as well as such extraordinary traits as nomadism, athletic ability, shiftlessness, and thalassophilia (love of the sea). The traits were, as with Galton's studies, diverse and by no means all heritable according to current thinking. But Davenport gathered an incredible data base of hundreds of family pedigrees encompassing at least three generations, obtained both from medical journals and from numerous family record forms which he mailed to various institutions, scientists, and associations.

Carnegie's funding was generous but Davenport's vision of an enormous central bureau of eugenics with information about the nation's protoplasm and the families that carried it required an even greater expenditure. Crucial to his success was a young woman named Mary Harriman. A graduate of Barnard College in biology, Mary spent part of the summer of 1905 working at the Cold Spring Harbor Biological Laboratory for Davenport. She was heiress to a railroad fortune and a social activist who later became a friend of Eleanor Roosevelt. Mary's exposure to Davenport persuaded her that his cause was worthy, and she arranged a meeting between him and her mother, Mrs. E. Harriman, who had assumed control of the family wealth after her father's death.

Davenport's appeal was carefully calculated; he knew Mrs. Harriman had been both daughter and wife to horsemen. An argument Davenport had used before that certainly was appropriate here was that they could bring about "the most progressive revolution in history" if only "human matings could be placed upon the same high plane as that of horse breeding."[10]

Harriman was convinced. She provided over half a million dollars in funding, spread over eight years, for the Eugenics Record Office, which became part of the biology laboratory at Cold Spring Harbor. This windfall provided seventy-five acres of land, a house, a fireproof safe (for the family pedigrees), and operating expenses including fees for field workers from colleges like Vassar, Radcliffe, Wellesley, Harvard, Cornell, Oberlin, and Johns Hopkins, who were dispersed around the country. Trait books in hand, this eager squadron collected hereditary information about inmates of institutions for the feebleminded, the insane, and the psychopathic as well as studying albinos and Amish. Detailed family histories were compiled on 750,000 cards in the Eugenics Record Office and kept for analysis. Davenport responded happily and authoritatively to many inquiries about the suitability of proposed marriages as well as issuing bulletins, books, and advice on sterilization measures, immigration policy, and eugenic theories.

The impact on society was palpable. As Davenport wrote to Mrs. Harriman, in words then warmly optimistic and now chillingly prophetic: "What a fire you have kindled! It is going to be a purifying *conflagration* some day!"[11] And so it was, especially in Germany, where *conflagration* was a horrifyingly apt description of what followed.

On the harmless side, there were Fitter Family contests held at state fairs, which sound humorously similar to the prize bullock and cherry pie competitions. However, the same sentiment fueled a widespread movement to begin scientific testing and assessment of human intelligence—seemingly an admirable aim but one that gave rise to a number of gross injustices. In 1904, the French government had asked psychologist Alfred Binet to concoct a means of measuring mental abilities, and he, with a colleague, Theodore Simon, obliged. The test—the direct ancestor of today's IQ tests—was brought to the United States soon afterward by Henry Goddard, who applied it to thousands of children, starting with those at the Vineland Training School for Feeble-Minded Boys and Girls in New Jersey. Goddard used the results to categorize the pupils into various degrees of feeblemindedness, ranging from idiots (those with a mental age of one or two) to morons (those who scored between eight and twelve). After studying a large family from the Pine Barrens of New Jersey, Goddard published an authoritative work on the inheritance of various undesirable traits called *The Kallikak Family: A Study in the Heredity of Feeblemindedness*.

The Kallikaks, hidden under a thin gauze of pseudonymity, seemed a perfect proof of eugenic principles. Like the Jukeses, the Kallikak family tree was so full of paupers, criminals, prostitutes, and depraved or feebleminded individuals that their name became a symbol of the undesirable. The Kallikaks could all be traced back to the bastard offspring of an upstanding man and a mentally deficient tavern girl. Since the same man later married a God-fearing Quaker woman and produced an admirable, law-abiding family, Goddard argued that it was the tavern girl's deficient genes that were to blame. Poverty, environment, social ostracism, or lack of education were given no credence as influencing the behavior of the Kallikaks; the family was seen as simply genetically and inherently evil. In fact, photos of the Kallikaks published by Goddard were deliberately altered to make the family appear sinister and diabolical.[12]

Under Goddard's enthusiastic promotion, Binet tests were also applied to potential immigrants at Ellis Island in 1913 to see how many mental defectives were being admitted. The testers were two women instructed to bypass the obviously normal; since government officials had already rejected the obviously defective, they believed their sample to represent the average immigrant, who was by these criteria neither normal nor defective. Most were tested in English (generally not their native language) and, of course, few could have had any idea what this bizarre exam was all about or whether cooperation would lead to denial of entry. Not surprisingly, an astounding 40–50 percent of those tested were ranked as feebleminded. Harsh measures were called for, Goddard concluded, even if a certain number of morons could be put to work at jobs too dull or routine for ordinary, intelligent folk to perform. He joined those lobbying for stricter immigration laws. In 1924, the Immigration Act was passed. In a blatant attempt to maintain the ethnic status quo, the new law dictated annual quotas. New immigrants from each country were limited to 2 percent of those of the same national origin that were living in the United States in the good old days of 1890.

Following rapidly on Goddard's heels were a series of earnest psychologists—the most famous being Lewis Terman and Robert Yerkes— who began mass testing, using an ever-evolving array of intelligence tests. Military recruits were an obvious, captive audience; an appalling number showed staggeringly low mental ages, with an average of about thirteen years. Negroes, Poles, and other peoples of eastern Europe placed even lower. Once again, that many of those being tested were illiterate, uneducated, or poorly fluent in English was glossed over by Yerkes and his staff. For his part, Terman was completely convinced that his modified test, known as the Stanford-Binet, actually measured intelligence. He was also certain that high intelligence was as closely correlated with morality and success as low intelligence was with criminality, so he urged universal testing for all children. The idea was that the tests would allow those in power to train each individual in tasks suitable to his or her ability; thus, the military could use the tests to select officers or those to be trained in skilled work and civilians could be aimed at professions to which their innate mental ability best suited them. Terman explicitly hoped to identify a minimum IQ for each occupation.

Put in these terms, the trend toward intelligence testing seemed beneficial or at least benign. But the other face of the Janus was the ugly visage of negative eugenics. Between 1907 and 1917, sterilization laws were enacted in sixteen states, starting with Indiana. These laws made it legal to sterilize criminals in particular categories as well as, often, epileptics, the insane, drug addicts, idiots, and—in Iowa—white slavers. The popular confusion between castration (removal of the testicles) and sterilization (surgically preventing the egg or sperm from reaching its partner) made it more common to sterilize those convicted of sexual crimes such as rape or moral degeneracy under the mistaken belief that sterilization would reduce sexual desires. To be sure, pungent critiques of sterilization laws were published. Charles A. Boston, a New York lawyer, argued that more people died annually of automobile accidents than were raped, so it would make more sense to sterilize "reckless chauffeurs"[13] than rapists. In 1919, Walter E. Ferdinald, an expert on feeblemindedness, had expressed doubts that mental deficiency was a single syndrome always attributable to genetic causes. These protests seemed to fall on deaf ears.

Although sterilizations slowed while the country was preoccupied with the First World War, the sentiments of fear and xenophobia were, if anything, exacerbated by the sense that the familiar world had come to an end. In 1927, a famous case known as *Buck v. Bell* went to the United States Supreme Court, over the question of sterilizing a seventeen-year-old "moral imbecile" called Carrie Buck. Both Carrie and her mother were certified feebleminded; now Carrie had given birth to an illegimate child named Vivian. The crucial question was whether or not Vivian was normal. Harry Laughlin of the Eugenics Record Office consulted the Bucks' pedigree and declared the child (at the time of the case only seven months old) to be feebleminded, like her mother and grandmother—a point confirmed by his colleague Arthur Estabrook, who tested the infant and also found her below average. The sterilization law of the state of Virginia was upheld by the Supreme Court, to the delight of eugenicists all over America. As Justice Oliver Wendell Holmes wrote:

> We have seen more than once that the public welfare may call
> upon the best citizens for their lives. It would be strange if it could
> not call upon those who already sap the strength of our being for
> these lesser sacrifices . . . in order to prevent our being swamped
> with incompetence. . . . Three generations of imbeciles are
> enough.[14]

Carrie's sterilization, thus endorsed by the highest court in the land, was performed. For good measure, in 1928, her sister Doris was also sterilized under the same law without being informed of the real reason for the operation. She was told it was an appendectomy.

As the Great Depression of the 1930s unfolded, public apprehension heightened. "Civilization is making the world safe for stupidity," warned Albert Wiggam,[15] a eugenicist speaking to an audience at the American Museum of Natural History. The tax burden of the ill, the insane, the indigent, and the criminal weighed heavily on the dwindling numbers of employed. In America, sterilization grew in popularity as a means of dealing with the habitually criminal, the poor, or the purportedly congenitally insane, in a twentieth-century version of transportation to Australia. Compulsory sterilization of those confined to institutions was entirely legal, and by 1941 as many as 36,000 had been operated upon.

However, an antisterilization movement grew apace with the numbers of sterilization advocates, especially in Britain which had, as yet, no sterilization laws. The basis for the concern was twofold. As data on feeblemindedness and related ills increased, it became apparent that perhaps as many as 90 percent of all mentally defective individuals were born to normal parents. These facts made it clear that a truly vast number of people of normal intelligence would have to be sterilized to remove these defective genes from the population altogether. Even this drastic measure would work only if mental deficiency has a purely genetic cause, a fallacy that was suspected by some in the first half of the twentieth century and would be exposed as such in years to come. The second point was that the specter of compulsory sterilization of the eugenically unfit loomed large. Even though the legislation proposed in England suggested that only voluntary sterilization be permitted—as in the case of the mentally deficient who were capable of living outside of institutions but who were not able to assume the responsibilities of parenthood—Parliament voted the bills down.

But in Germany, the view was rather different. The German people had inherited from Haeckel and other nineteenth-century thinkers a strong Romantic sense that fused fuzzy science with racism, contrasting the unwholesome and unwelcome traits of lesser races with the physical, moral, and intellectual glories of the true Germans: tall, blond, blue-eyed, clean-limbed country people of Aryan or Nordic descent. The Aryan race was a largely mythical creation, the people who spoke the proto-European language that was the common stock of Sanskrit, Zend, Armenian, Greek, Latin, Lithuanian, Slavonian, German, Celtic, English, French, Italian, Spanish, and so on. Physically, Aryans were assumed to represent an ancient and racially pure stock that carried this language, and their culture, from their place of origin near the Ganges into western Europe in prehistoric times. Protests, like that of Huxley in 1890—"community of language is no proof of unity of race,

[and] is not even presumptive evidence of racial identity"[16]—or Virchow's survey showing that most Germans were not blond-haired and blue-eyed had negligible impact on this widespread conviction. The Aryan or Nordic race existed and was the source of everything good, fine, and worthwhile in Europe and, for Germans, was closely identified with the mystical *Volk* of their homeland.

It was perfectly clear to many that Darwinian principles had led to the successes of the Aryans and their rise in German society; Haeckel had been saying so for years and, in 1906, founded a powerful organization of like-minded individuals called the Monist League. Membership in the League grew rapidly into the thousands and spread throughout Germany and Austria. Its meetings, lectures, and publications were only one form of influence; the League and fourteen other so-called Free Thought organizations joined together between 1907 and 1909 to form an anti-Church group known as the Weimar Kartel. Through the Kartel, the Monist League found a much broader audience for its racist propaganda and eugenics proposals. The League hammered home the message that the dilution of Aryan blood by inferior races was leading to the degeneration of the German, Aryan, or Nordic race. Social progress, the League proclaimed, was lagging far behind scientific progress, which was being hobbled by the ridiculous, antiquated notions propagated by the Church. It was time for a new plan, a different approach based on the scientific teachings of Darwin. Otherwise, the increasing numbers of unfit—already surely responsible for the paucity and poverty of Germany's overseas colonies, and for the financial and mental depression that swept the country after its defeat in World War I—would drag the greatest people in the world into oblivion.

Among Germany's foremost social Darwinists were Alfred Ploetz and Wilhelm Schallmayer. The latter was a member of Haeckel's Monist League and a pioneer in eugenics in Germany. Ploetz and Schallmayer did not argue for a laissez-faire economy, as had their counterparts in England and the United States. Instead, responding to a long-standing belief in Germany that the individual should be willingly sacrificed for the greater good of the state, they pushed for more government control, for intervention to halt and counteract the insidious downward slide of the German national and racial character. Ploetz, for example, argued that warfare, revolution, or government-sponsored medical care for the sick or unfit was "counterselection" that caused the deterioration of the race. It was he who coined the term *Rassenhygiene* (racial hygiene), a principle that would guide govern-

ment actions based on the good of the race, not the individual. Weak
elements must not be allowed to prosper and, above all, must not be allowed to reproduce.

In 1900, Friedrich Alfred Krupp, the prominent manufacturer of military arms, provided 30,000 reichsmarks as a prize for an essay contest. The topic was: "What can the theory of evolution tell us about
domestic political development and the legislation of the state?" The
judges were Haeckel himself and two members of the Monist League,
Heinrich Zeigler and Professor J. Conrad. The winner, Schallmayer—
like most of the other entrants, whose essays were published in a ten-
volume series—called for a careful and planned improvement of the
race. Eugenics was established in Germany.

Under Ploetz's leadership, the first steps were taken toward this improvement: the founding of a journal, *Archiv für Rassen- und
Gesellschaftsbiologie (Journal of Racial and Social Biology)*, in 1904
and, a year later, the founding of a society, the *Gesellschaft für Rassenhygiene* (Society for Racial Hygiene), that grew rapidly in membership,
especially after the end of the First World War. The first issue of the
journal was dedicated to two scientists whose work underlay the theory of *Rassenhygiene*: Ernst Haeckel, who brought Darwinism to Germany, and August Weismann, who provided the mechanism for
Darwinian evolution, by proposing that special germ cells in the body
carry and transmit the genetic information. The editors were almost
without exception members of the Monist League, and some would become prominent Nazi scientists. The journal was a major vehicle for
the dissemination of the eugenics ideas of Haeckel and his followers,
and after the rise of the Nazi party to power in the early 1930s the journal provided a respectable scientific showcase for Nazi views, which
differed little from the content of the journal in previous years.

When the Society for Racial Hygiene was formed, Haeckel and Weismann were declared honorary chairmen. They became, in effect, the
patron saints of eugenics in Germany, the geniuses who "proved" that
physical and moral character were inherited. Nature was all, nurture
nothing. By 1930, the society boasted 1,300 members across Germany,
with additional regional branches opening almost monthly. Members
had to swear to refrain from marriage if they were unfit but, as Robert
Proctor, a historian of science, has observed, there is "no evidence that
anyone in the society ever admitted such a defect."[17]

If there was no plasticity in human behavior and heredity was all-
important, then protecting and purifying the genetic material of the
race was an urgent priority. Any belief in the ability of humans to im-

prove themselves, through education, nourishment, or medical care, was seen as equivalent to a belief in the Lamarckian inheritance of acquired characteristics: a hopelessly old-fashioned and discredited point of view. A spate of publications that paralleled those in the English-speaking world argued that the unfit, the unemployable, the feebleminded, and the criminal were breeding far faster than the sound, the sane, and the hard-working. By 1931, when the Nazi party came into power, *Rassenhygiene* was their official policy. In 1937, the party published a coldly scientific document, the *Handbook for the Hitler Youth*, which was required reading for seven million young members. Among other topics, this book summarized Darwinian theory, Mendelian genetics, and Weismann's experiments refuting the inheritance of acquired characteristics.

> What we need to learn from these experiments [the handbook concluded] is the following: Environmental influences have never been known to bring about the formation of a new race. That is one more reason for our belief that a Jew remains a Jew, in Germany or in any other country. He can never change his race, even by centuries of residence among other people.[18]

Of course, the choice of the Jew as an example was no accident. The history of anti-Semitism in Europe in general, and Germany in particular, is long and dark. Since the Middle Ages, European Jews had been denied access to many professions, medicine being one of the few exceptions. Before the nineteenth century, they were legally prohibited in Germany from owning land or holding public office. The Antisemitische Volkspartei (Anti-Semitic People's Party) had existed in Germany since the late 1800s and in 1907 won twenty-one seats in the Reichstag.

Jews were a group who were commonly believed to be easily identified, though the official means of identifying who was Jewish in Germany would take on a Byzantine complexity. Relevant traits included details of facial features and the shape of the earlobes, but these were dissatisfyingly subjective. Great hopes were held out for blood types, discovered in 1900 by Viennese biologist Karl Landsteiner; as biochemical features identified in a laboratory by white-coated scientists, blood types seemed to be more accurate and closer to the genes themselves than the morphology or shape of anatomical features as a means of assessing either paternity and race. Blood type seems to work on a simple dominance scheme, with three alleles (or variants) of the gene:

A, B, and O. Both A and B are dominant to O; that is, one A or B gene masks the presence of the O. These combined to give four phenotypes (type A, type B, type O, and type AB) from six genotypes (AA and OA give type A; BB and OB give type B; AB gives AB; and OO gives O). The additional systems of blood typing used today, involving many more genes and alleles, such as the Rhesus and MN systems, were as yet unknown.

Genetic anthropologists, both in Germany and elsewhere, agreed that races and populations vary in their frequencies of genes for A, B, and O. In Germany, Otto Reche—a prominent professor of racial science—seized upon this fact as a possible means of identifying races. He believed that A was the blood type of northern Europeans (Aryans), B was originally Asian and Middle Eastern and was more common among Jews, and type O originated among pre-Columbian Indians in the Americas. (In fact, postwar surveys in Europe revealed that non-Jews had similar or somewhat higher frequencies of B genes than did Jews,[19] but race science was often more concerned with furthering political ideals than determining the truth.) In Reche's terms, A and B were "enemy" blood groups that should not be mingled, a metaphor that arose from the dangers of transfusing an A-type individual with B-type blood. It was not accidental that B genes also identified likely Jews, the enemy of the German state.

Jews were feared on account of their little-known and ill-understood religious practices, and were easily blamed for all of Germany's ills. Jews were not Aryan and they became, almost by definition, the universal scapegoat. As economic and social conditions worsened between the world wars, Jews came increasingly into focus as the acceptable target of mass dissatisfaction, chauvinism, and resentment.

Haeckel was a particularly outspoken anti-Semite, maintaining (among other outrageous beliefs) that Jesus Christ's father was actually a Roman officer who had seduced Mary, making Christ only half-Jewish; that Jews themselves provoked anti-Semitism, which was the only reasonable response to Jews' behavior; and that Jews not only avoided assimilation into German society but were actually biologically incapable of partaking fully of German culture. The influence of Haeckel's anti-Semitic views on German society and the Nazi party was immense because of his huge personal following and high scientific standing. Indeed, the sound of Haeckel's words almost certainly rang in the ears of Adolf Hitler himself. Historian of science Daniel Gasman has argued that the substance and wording of some of Hitler's writing "emerge as an extended paraphrase and at times even plagiarism of Haeckel's *Natürliche Schöpfungsgeschichte* and the *Welt-räthsel,*"[20] Haeckel's two most popular works. Hitler even shared Haeckel's fondness for the designation *Kampf*.[21] It is only reasonable to suppose that Hitler did read these books: certainly on occasion, Hitler referred directly to many of Haeckel's most important ideas, including the biological unfitness of Jews and the sure doom that would befall the German people if they did not cleanse themselves of such impurities. At a eulogy written in 1934 for the celebration of the centenary of Haeckel's birth, Professor Gerhard Heberer proclaimed, "It is to be recalled at this opportunity that Haeckel was one of the first fighters for eugenic measures. His proposals were being brought into reality in the new Reich."[22]

And they were. In the years between the wars, racial hygiene became institutionalized in Germany, with the blessing and encouragement of the highest levels of government and science. In 1927, the Kaiser Wilhelm Institute for Anthropology, Human Genetics, and Eugenics was founded. Its aim was to carry out research that would aid the struggle to restore racial purity and strength to the German people. Like eugenics organizations in the States or in England, this office conducted massive surveys on the incidence and inheritance of a wide range of ailments; the scientists employed at the institute also helped draft legislation regulating marriage laws. The first director of the institute, Eugen Fischer, was apparently a little too moderate for the Nazi party. In 1933, he was warned obliquely of the party's view, being replaced as head of the Society for Racial Hygiene by Ernst Ruden and subsequently denounced by various other scientists. Fischer responded to this clear threat by allying himself more closely with mainstream Nazi doctrine.

But it is not the whole truth to say that the Nazi party forced racial hygiene upon the biomedical and anthropological scientists of the day. It must be remembered that it was precisely this community which invented racial hygiene—a point made most convincingly by Robert Proctor in his book *Racial Hygiene: Medicine under the Nazis*. Coercion was generally applied by scientists of one school of thought to those of another; Nazi thinking was not simply foisted on innocent and reluctant scientists by evil politicians. The roots of the Nazi racial policies reach back to Haeckel on scientific matters and even farther into history in terms of social attitudes like anti-Semitism and xenophobia.

As the Nazis' power increased during the 1930s, the amount of legislation dealing with eugenics and racial hygiene increased dramatically and insidiously. Institutions, offices, and policies sprouted as vigorously as weeds in a neglected garden; they were not the products of neglect but of earnest cultivation. Starting in 1931, members of the elite SS were prohibited from marrying non-Aryans and SS Chief Heinrich Himmler established a Racial Office of the SS to guarantee that all marriages and sexual liaisons conformed to Nazi racial doctrine. In 1934, an Office of Racial Policy was established by the Nazi state, which was to "coordinate and unify all schooling and propaganda in the areas of population and racial policy."[23] Publications, meetings, and films poured out of the Office of Racial Policy, all designed to "enlighten" the public about the dangers of the inferior races, especially the Jews. New legislation systematically barred Jews from employment in the civil services, including university professorships and many medical positions.

The large number of Jews in the medical profession was a particular concern to Hitler and the Third Reich since a distorted brand of "Darwinian" biology provided the intellectual underpinnings of so many of their policies. While Jews were less than 1 percent of the German population in 1933, they constituted 13 percent of the physicians, because this profession had been one of the few open to them traditionally.

The Nazis believed it was dangerous to give Jews such power and "dominance" over the health and breeding success of the Aryan population, which was a major priority. To encourage the birth of Aryan children, Hitler established the Honor Cross of German Motherhood in bronze (four children), silver (six children), and gold (eight children), which was given to thousands or perhaps even millions of women. Large loans were available to men whose wives gave up jobs outside the home and the principal owed on the loan was diminished by 25 percent at the birth of each child.[24] Women were strongly discouraged and, in some instances, prohibited from most types of work or profession outside the home. It was seen as contradictory to glorify motherhood as the sacred duty of German women and yet to entrust the health of the women and children of Germany to the hands of Jews.

The expulsion of Jews from the medical profession began early in the Third Reich. Successive government measures were passed that meant that the fees of Jewish physicians were no longer reimbursable under the government medical insurance program, and in 1935 the entire medical profession was subordinated to the government. Many Jews were simply eliminated from the profession by legislation and harrassment that continued to gain in stridency until the Fourth Ordinance of the Citizenship Law was passed in 1938, revoking all medical licenses issued to Jews. Jewish physicians could treat only other Jews, with special permission, although a select few would be permitted to work as attendants to Aryan physicians to avoid overburdening the Aryans. Large numbers of jobs in medical schools, clinics, hospitals, and institutes were freed up for Aryan physicians, many of whom had been out of work and, according to some authorities, even starving during the worst part of the depression. This much-needed financial relief obviously prevented many Aryan physicians from protesting the anti-Semitic measures.[25]

Interspersed with legislation designed to remove Jews from the medical profession were other anti-Semitic laws, such as that requiring registration of all Jews, the law banning Jews from a wide range of professions, and the order prohibiting Jews from attending German theater, concerts, lectures, cabarets, circuses, variety shows, dances, or cultural exhibitions or attending German schools. It was a systematic, deliberate, cruel, and vicious discrimination against the Jews of Germany that nonetheless contained many of the elements of eugenic thinking that were commonplace in Britain and the United States. One important difference was the identification in Germany of a single group, the Jews, as the primary bearer of unfitness. Another was the disturbing ease with which discrimination against this group *on purportedly biological grounds* was accepted.

The frequency of the passage of new laws designed to enhance the racial purity of Germany accelerated through the 1930s. The Nuremberg Laws, enacted in the fall of 1935, made marital counseling obligatory—meaning that those intending to wed had to be examined and certified as racially and physically suitable for marriage, upon threat of imprisonment for those who wed in defiance of the advisory center's decision. These laws were widely perceived as simple public health measures, like the familiar blood tests for venereal disease required for marriage licenses in many parts of the United States today or the prohibition against typhoid carriers working in the food industry. It was part of a broad program. As early as 1933, the Law for the Prevention of Genetically Diseased Offspring had been in effect. Under its terms, an individual could be sterilized on eugenic grounds, for example, if he or she were judged to suffer from genetic illnesses, which included epilepsy, Huntington's chorea, feeblemindedness, manic-depressive insanity, genetic blindness or deafness, schizophrenia, or severe alcoholism. Such sterilizations were compulsory, and the law was administered through 181 genetic health courts, whose workings and records were confidential.

At the same time, public awareness of the economic burden of "useless lives" was increased through an effective propaganda campaign. Books and movies portrayed euthanasia of the mentally or physically defective as a kindness and a release from suffering. Posters showed the stalwart German worker staggering under the immense "load" of the genetically ill, who were unsavory characters balanced on a beam laid across his shoulders, in an image evocative of those of Jesus Christ carrying his cross to Calvary.[26] Schoolchildren were set mathematics problems calculating the cost of caring for the mentally ill, being asked, for example, how many much-needed housing units could be constructed for good German workers for the same expenditure.[27] At its heart, the Nazi's euthanasia policy was based on economics and on the prevailing sense that times were difficult; defectives were expensive. The cunning part of the strategy was that it was passed off as a simple medical solution to an epidemiological problem.

The systematic killing began in 1939 with retarded or handicapped children under the age of three who were confined to institutions. They died by injections, poisoning, gassing, slow starvation, or even exposure when institutions were not heated, but they died mostly at the hands of medical professionals. Most sinister of all is the fact that these physicians and nurses were not *ordered* to kill patients; they were simply *empowered* to do so. The killing soon spread to adult psychiatric patients; some 70,273 were killed by 1941. Rumors of the killings grew to the point that some elderly refused to enter retirement homes, fearing those would be the next target.[28]

But the more obvious target was the Jewish community. Once the link was made between those who were genetically defective (in the sense of bearing physical or mental handicaps) and euthanasia, it was a simple matter to extend the principle to justify elimination of the Jews. In the views of Nazi physicians, Jews were pathological—a diseased race. "There is a resemblance between Jews and tubercle bacilli," one physician wrote in a ghastly rationale. "Nearly everyone harbors tubercle bacilli, and nearly every people of the earth harbors the Jews; furthermore, an infection can only be cured with difficulty."[29] In short, the Jewish "problem" became a medical problem, and the Final Solution was cast in those terms, as were myriad laws that prohibited Jews from intermarrying with Aryans or frequenting public places, where they might "infect" the German populace. Jews had always tended to live in specific areas; starting in 1939, they were confined by law to ghettos—on "hygienic" grounds. In some cases, as in Warsaw, Poland, walls were built around the perimeter of the ghettos to keep the Jews within, in what were called quarantine measures. Forceable resettling of Jews from rural areas into these ghettos overstretched water supplies and sanitation measures, leading to epidemics of typhoid, tuberculosis, and other crowd diseases that simply seemed to confirm the allegations that Jews were dirty and diseased. The government created the problem, heightened its urgency, cast it in medical terms, and then pressed the medical profession to find a Final Solution.

One proposed solution was sterilization. Viktor Brack proposed to sterilize all Jews capable of work—the others were to be eliminated—by having them stand at a counter to fill out a form and irradiating them without their knowledge. But because gas chambers for killing the mentally ill already existed, and had been proven to work so well and with so little fuss, gassing Jews became the method of choice. It began with Jews in psychiatric hospitals and was expanded, in 1941, to include all Jews in hospitals for any reason. Jews were separated from mainstream society into concentration camps, on the ground of public health. The able-bodied were forced to work for the benefit of the state, and the murder of those who were infirm, exhausted, or ill, or of those who were not Jewish but were politically or "socially" dangerous, was an obvious next step in the Solution and was taken in late 1941 or early 1942. No distinction was made between prisoners eliminated on purely medical grounds and those selected for termination for political or racial reasons, for the simple reason that these were all regarded as a single category of contamination. Jews, homosexuals, Gypsies, communists, prostitutes, and those suffering from tuberculosis, mental illness, retardation, deafness, or a wide range of other problems were selected for death. Although there were protests against these policies, they were too ineffectual to make any significant difference.

Beyond the sheer horror of the mass killings were the brutal "medical experiments" carried out on some prisoners. The flimsy rationale was that all prisoners were effectively marked for death anyway and that experimentation might lead to improved survival for Nazi soldiers or improvements in the health of the good German people. Natural selection and evolutionary theory, warped by the powerful lens of Nazi beliefs, made clear that the survival of the fittest was all that was important.

These policies were a shameful and terrible distortion of Darwin's ideas that moved the rest of the world to horror and violent protest. There was a widely shared sense, during the Second World War, that the issues were simple ones of good versus evil—a justifiable reading of the situation that lent a certainty to the war effort that has not been duplicated in any major war since. In a fundamental sense, the Final

Solution and the dreadful maiming of evolutionary theory that supported it were evil.

What is most appalling is the close resemblance of popular sentiments in Germany and in Allied Europe—the frustration with economic hardship and the pointed resentment of the unfit as a burden responsible for those troubles. The eugenics movement thrived in England and the United States, as well as many other European nations, for the same reasons that it attained mass popularity in Germany. Involuntary sterilization was not confined to Germany; it was practiced in the United States, Canada, Denmark, Finland, Sweden, Norway, and Iceland, too. In fact, euthanasia of the unfit or the hopelessly ill was proposed in many more countries. Another example is legislation against "mixed marriages." As late as 1942, thirty of the United States had laws on the books forbidding miscegenation of various types.[30] Prohibited marriages include those between "whites" and "Negroes" (variously defined by the percentage of Negro blood, by African descent, or by the number of generations since a Negro), Mongolians, Indians, Hindus, Chinese, Japanese, Ethiopians, Koreans, Kanaks, mulattos, mestizos, half-breeds, colored persons, and Malays. Disturbing as these laws and beliefs are from today's perspective, they fell well short of government-sanctioned plans to exterminate whoever is designated as inferior.

Across the Western world, the social context and commonness of prejudicial beliefs were similar to those in Germany, where Darwin's benign theory was subverted into an instrument of blatant and wicked genocide. Fortunately, conditions were just sufficiently different to prevent growth of the same noxious result elsewhere, even if eugenics in a milder but still insidious form flourished temporarily.

As the full truth of the Final Solution in Germany reached the outside world, the eugenics and sterilization movements withered. The hollowness and unsoundness of the principles of eugenics and of concerns for the evolutionary well-being of the human race were suddenly revealed. Eugenicists recognized, all too well, the similarities between their arguments and those of Nazi Germany; they saw, with bitter clarity, the dreadful reality they so closely approached, not knowing it was there, and they recoiled from it.

But remnants of the movement persisted. There were still those who believed that understanding the genetic inheritance of abilities or tendencies toward illness was important and would provide guidance for improving society. This group divorced itself utterly from negative eugenics and its penchant for manipulation or even subordination of the populace; they were among the loudest crying "Never again." Yet attempts to apply Darwinian evolution to humans did not die. The biologists who believed that the study of human evolution might bring improvement migrated into the more respectable field of human genetics.

Part IV

The Genetics
of Evolution

As Blond as Hitler

CHAPTER 8

AN INTEREST IN HUMAN GENETICS did not make a scientist a Nazi. A new breed of biologist shrank back from eugenics and racism, repelled, when the horrors of Nazi Germany came to light.[1] In reaction, a coterie of scientists formed who believed in the power of modern science to do good as well as evil. And as the Allied nations drew together to fight both the Nazis and their misuse of Darwinian thought, the scientific communities of Britain and the United States began to communicate and collaborate, driven by adversity, desperation, and a common language. If the war was destroying the old world, then it was time to create a new and better one using the potent new tools that science provided.

Human evolution had been politicized in Germany as early as the first introduction of Darwin's words into that nation. Haeckel and Virchow had, for their separate reasons, continued this tradition, infusing evolutionary theory with the burning questions of the day concerning racial identity and ancestry. But race science, as some called it, took on a momentum of its own, gathering speed and power as it was appropriated by the Nazi party. Medical practitioners and physical anthropologists became race scientists under pressure from the Nazi government, transforming warped beliefs about race into a terrible implementation of "public hygiene measures" that marched, inexorably, toward ultimate evil. An explicitly recognized part of this "sanitation" program was to issue an overwhelming flood of racist propaganda—in films, books, articles, lectures, posters, and radio programs.

The obvious response was for politically minded scientists to issue their own, antiracist propaganda, making the scientific truths about racial differences known. This act was the tangible manifestation of a growing unease with some of the methods and conclusions of scientific attempts to deal with human races. It was time for combat Darwinism—Darwinism as interpreted by the political liberals, used as a weapon for good.

Though anthropologists had been measuring skulls with fanatic enthusiasm for years—convinced that cranial form was a stable, heritable indicator of race—the results were unsatisfactory. In a notable example at the turn of the century, William Z. Ripley was compiling his book *Races of Europe* and he wrote to Otto Ammon, who had conducted an extensive anthropometric study in Europe, for a photograph of a "pure Alpine type" that he had described from the Black Forest of Germany. Ammon was unable to oblige. "He has measured thousands of heads," marveled Ripley, "and yet he answered that he really had not been able to find a perfect specimen in all details. All his round-headed men were either blond, tall, or narrow-nosed, or something else they ought not to be."[2] Something was wrong, with either the methods or the conclusions.

Still, science seemed the appropriate weapon to take up against pseudoscience. Once again, a Huxley became a moving force, guiding the development and application of evolutionary theory to humankind. Julian Huxley, Thomas Henry's grandson, and Alfred C. Haddon, an anthropologist formerly at Cambridge, together wrote one of the most influential, antiracist books of the era. Published in 1935, it was called *We Europeans: A Survey of "Racial" Problems*. It was written with all the style and skill that this Huxley could muster and all the credibility and depth of knowledge that Haddon could command.

Julian Huxley was his much like his grandfather in temperament and intelligence. "I like that chap," Thomas Henry Huxley said on one occasion, referring to his high-spirited four-year-old grandson. "I like the way he looks you straight in the face and disobeys you. I told him not to go on the wet grass again. He just looked up boldly, straight at me, as much as to say, 'What do *you* mean by ordering me about?' and deliberately walked on to the grass."[3] In another family, such a precociously bright, independent, and challenging child would not have thrived, but the Huxleys always prized cleverness and originality and set no store by conventionality. By the time he was a student at Cambridge, Julian's intellectual gifts and creativity were well evident.

This Huxley was tall, quick-witted, a keen observer, and an original thinker. He found racism as worthy a foe in his day as the Church had been in Thomas Henry's. Over the next few decades, Julian would have an enormous impact on world opinion about evolutionary and racial matters. As the words of Thomas Henry Huxley indelibly colored the understanding of Darwinian evolution in the nineteenth century, so would Julian Huxley's tint human evolution in the twentieth.

We Europeans was not his first turn in the limelight, however. Back in 1920, as a young biologist, Julian Huxley had caught the attention of the popular press with an experiment that seemed to prove the existence of an "elixir of life" capable of creating a new animal. What Huxley had discovered was, in truth, a thyroid extract, and his experiment was one of the earliest demonstrations of the dramatic effect hormones have in controlling growth and maturation. The occasion demonstrated that this Huxley was as pungent, clear, and effective at explaining science to the public as his grandfather had been. From then on, Huxley was established as one of the foremost scientists who could explain biology to the ordinary people of Britain, in ways they found fascinating and comprehensible. By the 1930s, Huxley had a sort of celebrity-scientist status, which he used to the fullest in a tremendously popular radio series that ranged from natural history through the latest scientific research in embryology to the role of science in fostering a better society.

It was this last theme that Huxley and Haddon's book dwelt on. They enjoyed a built-in audience, a guaranteed public who already listened regularly to Huxley's voice on the radio. Knowing the insidious mixture of half-truths, lies, and fears that the Nazis were disseminating, the pair attacked "race science" directly. They used modern genetics to highlight flaws in works like Madison Grant's *The Passing of the Great Race*, read by thousands in England and America, and by implication to render ineffectual even more inflammatory Nazi propaganda. Their argument was all the more effective because of the full weight of Huxley's public and scientific influence.

One of their most important points was that pure human races were imaginary constructs; they did not and could not exist, and thus could not have consistent behavioral or temperamental characteristics. This is why it was impossible even to define the races consistently, despite the exercise of much ingenuity. Even in Hitler's Germany, it was proved difficult to figure out who was and who wasn't Jewish, leading to an obsessive concern with genealogies.

Haddon and Huxley pointed out that even the most modern and objective scientific techniques foundered on the problem of racial classifications. Skin and hair color, or hair type, had been shown by Virchow to be tremendously variable and unreliable. While any fool could look at a series of photographs taken in the cities of Oslo, Beijing, and Lagos and see that the inhabitants look different, it is a complex statistical phenomenon; the most obviously varying traits are not linked to each other at a genetic level. Dark eyes may occur with fair hair, aquiline noses with broad cheekbones, tall stature with the epicanthic eye fold often described as being Mongoloid, and so on. The fossil record, which was much better known by the middle of the twentieth century, attested to the long-standing human propensity to migrate and interbreed. Racial groupings based on blood types produced one classification of the peoples of the world, while those based on measurements of head shape yielded another, and those relying on linguistics came up with a third. More to the point, Haddon and Huxley asserted authoritatively that intelligence, moral worth, and other character traits were not reliably linked to any of the so-called races the Nazis and others recognized; these were more properly labeled ethnic groups in any case. Perhaps the notorious Huxley family wit came into play, too, for the most famous passage in the book ridiculed Nazi stereotypes by constructing a true Aryan or Teutonic type from the political and ideological leaders of Nazi Germany: "as blond as Hitler, as dolichocephalic [long-headed] as Rosenberg, as tall as Goebbels, as slender as Goering, and as manly as Streicher."[4] Satire, as several Huxleys have shown, is a potent weapon.

Huxley, with his wide-ranging interests that embraced both the new molecular aspects of biology and the older tastes of field naturalists, was something of an anomaly as a biologist between the wars. Like his social conscience, his eclecticism thrust him into a prominent role during this era.

By the time World War II was declared, biology had reached a curious impasse.[5] Darwin's major stumbling block—how traits were inherited—had been to a large extent overcome. By the early years of the twentieth century, the broad principles of genetics and inheritance had been established by men like August Weismann, who had proposed that there were germ cells separate from the ordinary bodily cells—germ cells that carried the genetic material, which was recombined and assorted as the mother's egg joined with the father's sperm to form a new being. Another source of insight was the rediscovery of the meticulous studies of an Austrian monk, Gregor Mendel, whose now-famous research on inheritance in pea plants had lain largely unnoticed since 1866. Within the span of a few months in 1900, three botanists—the

Dutchman Hugo de Vries, the German Carl Correns, and, to a lesser degree, the Austrian Erich Tschermak—independently worked out the mechanics of genetic inheritance. To their surprise, they subsequently discovered that Mendel had anticipated them by decades, albeit in rather scanty and obscure publications. In 1909 the term *gene* was suggested as a word to designate the physical basis of an inherited character, in recognition of the distinction between the expression of traits (phenotype) and the underlying units of inheritance for those traits (genotype). But the diverse implications of the inheritance of traits split biology into two distinct camps, which rarely communicated with each other and which shared few scientists in common. Only a few oddballs, like Huxley, stayed in touch with both parts of modern biology and chose to enlist their science in the service of peace and human understanding.

One camp was natural history—that genteel, vaguely defined, and broadly inclusive field of study—which incorporated zoology and botany, paleontology and comparative anatomy, inheritance and variation, geographic distribution and ecological adaptation. While these areas of inquiry were rarely mastered by a single individual, all in this camp were, in a broad sense, field biologists: people who studied organisms in the wild or, in the case of physical anthropologists and paleontologists, in the fossil record.

The other camp was the burgeoning discipline of experimental biology or genetics. Experimental biology was divorced from the Darwinian type of field work and focused on the laboratory and the notebook. Scientists in this camp wanted to understand the details of inheritance and how traits spread through populations; their emphasis was often on numbers.

Genetics was a term coined by William Bateson, who was acutely sensitive to the importance of Mendel's laws of inheritance. Eugenicists like Karl Pearson had provided many of the statistical and mathematical tools for analyzing large amounts of data on inheritance; now Bateson and others began to breed, study, and manipulate large populations of experimental animals, focusing on creatures such as the marine "worm" *Balanoglossus*. These animals showed readily recognizable and quantifiable traits that were passed on from generation to generation, each arriving rapidly after the last. In the United States, Thomas Hunt Morgan set up a famous, collaborative genetics group working on the ideal experimental animal: the rapidly breeding fruit fly, *Drosophila*. He and his colleagues became leaders in understanding the roles of mutation, variation, and sex linkage in genetics.

Some of these experimental biologists were more properly called mathematical geneticists. They really hoped to understand *human* genetics in populations—men like Ronald A. Fisher and J. B. S. Haldane in Britain and Sewall Wright in the United States. Britons dominated the field up through the 1940s, receiving a tremendous boost from wartime medicine because of the risks and advantages of blood transfusions. Documenting and understanding blood-group genetics was of obvious practical and humanitarian value, and a group of scientists centered in London attacked the problem with enthusiasm. For those unfit to serve in the armed forces—Fisher, for example, was so nearsighted that he was literally capable of knocking his pipe ashes into the butter[6]—such work gave them an important sense of contributing to the war effort.

To some extent, these new biologists fit the stereotype of the true English boffin: an esoteric, absentminded, unworldly scientist. Fisher, for example, was not only nearsighted, he was in other senses a misfit, ill at ease with women or strangers, unable or unwilling to finish his undergraduate degree at Cambridge, and on occasion downright rude. He was also openly a eugenicist and a close friend of Darwin's son, Leonard, the longtime president of the Eugenics Education Society, from their undergraduate days. Similarly, Haldane was painfully shy— a trait he concealed at times behind inappropriately vulgar remarks— and utterly inept at manual tasks. He was no man of action, no field biologist with mud on his boots and specimens in his back garden; he was more at home with numbers and ideas than with living creatures. But now that the mechanisms of inheritance had been worked out, it was time for some abstract modeling of how genotypes and phenotypes might change over time in a population—work for which men like Fisher or Haldane were eminently suited. In the United States, at the University of Chicago, Sewall Wright was a man of similar tastes. He combined experimental breeding of guinea pigs with the work for which he is more generally remembered: complex mathematical models of inheritance in dynamic populations.

These men were wrestling with an urgent intellectual problem: how exactly did the genotypes of a population change over time? Was it possible for small, advantageous variations to sweep through an entire population or species and change it into something new? If not, Darwinian evolution was indeed dead, as some were asserting. Their affirmative answers, derived from painstaking hand calculations, brought new insights into the way evolution might occur.

But what should have been a confirmation of Darwin's accuracy backfired. Rather than pointing to natural selection as the main or only motor producing evolutionary change, the geneticists became entranced with a concept known as *genetic drift*. Sewall Wright coined the term to refer to a phenomenon he observed in his modeling experiments: random fluctuations in reproductive success could dramatically alter the proportions of different genotypes within a population over the course of generations. That is, a population could *drift*, by accident, into an entirely novel genotype and phenotype. An example of drift can be envisioned in a small population of humans, such as colonists in a new land. If redheads, initially uncommon, happen to bear more children than brunettes for a few generations, the frequencies of hair-color genes in the entire population might be radically altered. Indeed, in a small, founding population, the chance failure of a single individual to breed (or to raise his or her children to reproductive age) may have long-lasting effects on the frequencies of that person's genetic traits. In time, purely through chance, redheads might come to predominate.

Wright found genetic drift to be especially potent in small, isolated populations; his work explained why such groups sometimes acquired special characteristics that are rare or even unknown in the main population of the same species. For their part, the experimental geneticists emphasized the crucial role of mutation in producing evolutionary novelties; they, too, tended to dismiss natural selection—Darwin's mainstay—as irrelevant.

During this period, the confidence in the power of science for good was tremendous, and scientists did not shrink from advocating bold steps to improve the world. In 1924, J. B. S. Haldane made a radical suggestion in a book called *Daedalus*, in which he outlined a utopian vision of the future controlled by genetic engineers. The mythological Daedalus arranged the creation of the monstrous Minotaur, by mating Pasiphaë with the Cretan bull; Haldane proposed a more benign arrangement. The new Daedalus, Haldane's Daedalus, would free human reproduction from the messy business of sexual love altogther. Those whose genes would be propagated would be selected on a scientific basis. "Ectogenic children" would be produced in laboratories, and the world would improve by leaps and bounds: "the advance in each generation in any single respect, from the increased output of first-class music to the decreased convictions for theft, [will be] very startling,"[7] he opined.

A similar suggestion came from Hermann Muller, a son of German refugees who joined Thomas Hunt Morgan's group at Columbia and then left the United States for the Soviet Union. Muller and Herbert Brewer, a eugenicist and socialistic post-office clerk in Britain, wanted to promote "eutelegenesis"—what would now be called artificial insemination, but with eugenically selected human stock—as spelled out in Muller's book *Out of the Night*. It was amazingly popular. Haldane, apparently pained by his own childless marriage, volunteered to supply his "name, money and gametes"[8] to eutelegenesis; George Bernard Shaw and Julian Huxley were also taken with the idea, at least in the abstract. But in contrast, Aldous Huxley—Julian's brother and a gifted writer—responded cynically to these suggestions for genetic engineering. In short order, he wrote *Brave New World*, a novel in which he portrayed the sterile, heartless society that such a scientific utopia might actually become. In another work, Aldous Huxley tweaked Haldane himself with a recognizable, satirical portrait of a boffin too preoccupied with obscure experiments to notice his wife's adultery.

While these highly publicized suggestions for human breeding captured the imagination, such issues were remote from the real focus of the geneticists. Their thoughts revolved around numbers and shifting percentages; what the traits actually were or looked like was unimportant, as long as the fruit fly (or guinea pig, mouse, or other creature) could be accurately scored. The mathematical geneticists were even further afield and never dealt with actual traits or living creatures. Theirs was an abstract study, highly theoretical and divorced from the squawling realities of real animals eating, breeding, and evolving in real environments. The genes or traits that were the genuine object of interest might almost have been disembodied. The genotype was all, phenotype nothing. Natural selection was of minor interest, adjudged far less powerful than genetic drift or mutation in evolutionary change.

While this branch of biology struggled to grasp genetics and the details of the evolutionary process, the other branch remained fascinated with field biology and held true to the Darwinian concept of natural selection. Prime among them were taxonomists, whose daily concerns were with variation, classification, adaptation, and geographic distribution of species. Paleontologists and comparative anatomists focused on the same sort of issues, but their species were spread out over an inconceivably longer time-span: millions of years, not the weeks, months, or few years of those who studied living species. Their hands were, metaphorically, full of visible characters—length of beak, angle of jaw, or number of teeth or toes—and not the hidden genes whose presence was manifested in these characters. They were tracking the evolution of their species through time and space, using as spoor the anatomical evidence of the animals' bodies. Their data were purely phenotypic. Mendelian genetics was not only invisible, it was all but irrelevant to the naturalists.

In short, the naturalists and geneticists never spoke, never read each other's papers, because their concerns and their raw materials were totally disparate. The two camps also held views of the evolutionary process that appeared to be as irreconcilable as if they were thinking and writing about entirely different effects—which, of course, they were. But suddenly, in the late 1930s and early 1940s, the stalemate broke almost magically and the tension between these two subfields gave rise to a new, fertile burst of energy.

Exactly how and why the new evolutionary synthesis came about has been the subject of much discussion. It is clear that, first, some willingness for each type of biologist to enter into a dialogue with the other was needed; as naturalists began to think about the subjects of their field studies as evolving populations and geneticists began to consider real, not just theoretical, populations, a rapprochement became easier. Too, some prominent individuals with the temperament and training to be bridge builders were needed: among those often singled out for this role are Julian Huxley, Theodosius Dobzhansky, George Gaylord Simpson, and Ernst Mayr.

What united these scientists? One obvious trait was their breadth of training and interest. Huxley was already accustomed to the public role of spokesman and synthesizer; his personal research was eclectic, ranging from laboratory biology to field studies of bird populations. His book *Evolution: The Modern Synthesis*, published in 1942, became a widely read statement of the way in which genetic theory illuminated the classical problems of speciation, variation, and evolutionary change in the field. Dobzhansky, a Russian refugee, could be cast as a classical experimental geneticist, but for the fact that he coupled his laboratory work with studies of free-living populations of *Drosophila*. His 1937 book, *Genetics and the Origin of Species*, marked the beginning of the synthetic process. In it, he openly attempted to unify the new, theoretical genetics with a Darwinian concern with real populations of living species; it was an influential and masterful integration of the two disparate ends of biology. G. G. Simpson brought another discipline into the synthesis; he was a well-known, broadly trained paleontologist, one of the few who looked at the fossil record in terms of evolutionary trends and tendencies. In a series of books beginning in 1944 with *Tempo and Mode in Evolution*, Simpson demonstrated that the genetic principles of change in populations were visible in the fossil record, the only evidence available that was on an appropriately long time scale. The fourth architect of the new synthesis was Ernst Mayr, a renowned bird taxonomist and field biologist. He, too, wrote a book with a far-reaching influence, *Systematics and the Origin of Species*, that brought the newly elucidated mechanisms of inheritance, mutation, and genetic drift to bear on the naturalist's problems with taxonomy, adaptation, and speciation.

Mayr's proposal for a new definition of the word *species* was virtually universally adopted; it is a revealing concept that demonstrates the effectiveness of the synthesis that was taking place. Instead of recognizing a species in the old-fashioned, traditional mode—by matching various physical characteristics of the unknown individual to those of some preestablished archetype—Mayr proposed the *biological species concept*. He defined a species as a group of potentially or theoretically interbreeding individuals who were reproductively isolated from other such groups. In other words, a species was not a medium-sized, predominantly blue-feathered bird with a white eye-stripe and black bars on its wings; a species was a circumscribed collection of individuals who exchanged and recombined their genetic material in a single gene pool. Mayr argued that what mattered about a species, in the evolutionary sense, was its genotype, not its phenotype; reproducing fertile offspring was the ultimate issue in the evolutionary theater and the action was at the species level. It was a new and empowering perspective on old problems. His definition has remained central to modern evolutionary biology to the present.

There were, of course, many others who contributed to the revitalizing synthesis of genetics and Darwinian evolutionary theory, creating a consensus and fusion where once there had been only dissent and confusion. Mayr, in his several writings about this period,[9] points to a landmark conference, held at Princeton, New Jersey, on January 2–4, in 1947, as the moment of truth. As he and others recall, there was an unanticipated and unforced agreement on this occasion, despite the fact that the participants represented diverse groups of specialists. Suddenly, all concurred that evolution was most often a gradual process driven primarily by natural selection, though genetic drift and mutation were accorded roles, too; suddenly, it was evident that the large-scale, visible evolutionary changes that gave rise to speciation events were fueled by shifts in genotype frequencies in smaller populations. Darwinism was empowered by the rich new motor of genetic mechanisms; genetics and population biology were, for their part, newly rooted in the nitty-gritty of real-world biology, of animals and plants that lived and bred.

It was time to turn back to the knotty problem of the value of the differences among humans.

All Non-Jews
Are Anti-Semitic

CHAPTER 9

W ITH THE SCIENTIFIC TENSION between the two branches of evolutionary biology resolved, it was time to look forward to the postwar peacetime and how the new evolutionary science could be used. It need not be an agent of darkness and evil.

Julian Huxley came to play a vital role in this new trend, again exercising his social conscience and passionate convictions. Even while the war seemed blackest, plans were being laid to establish an international organization—what would become the United Nations—to mediate and resolve international conflicts and to coordinate international efforts. But this was not enough. In 1945, a special agency based in Paris was set up to deal with education, science, and culture: the United Nations Educational, Scientific and Cultural Organization, or UNESCO for short.

The first executive secretary of this fledgling organization with a rather poorly defined agenda was Sir Alfred Zimmern, appointed in 1946. Unfortunately, he fell seriously ill and another executive secretary was sought. Julian Huxley, at first reluctant, was persuaded to take on this nebulous but important task.

There was no clear mandate for UNESCO and no infrastructure whatsoever. Huxley had to define his organization's job and build the mechanism to accomplish it simultaneously. His enthusiasm and ability to inspire loyalty in others were major assets; so too was his Huxleyan eclecticism. "I must confess it has been rather bewildering," he remarked in the early days, "especially for the deputies and myself, who find ourselves jumping from fine arts to applied science, from applied science to architecture, from architecture to rural education, from rural education to literature and philosophy—all in the space of a few hours."[1] Intellectually, he was perfectly suited to the job.

What Huxley lacked was any appetite for the daily minutiae of administration, for the endless plans and records and approvals. In some sense this was an asset, too, for he followed his instincts and beliefs directly, with little consideration for their practical consequences. One of his first acts was to write a document that expressed his personal view of the mission and direction for UNESCO. No one else wrote a word in it, no one collaborated in it, approved it, or censured it. *Unesco: Its Purposes and Philosophy* has been called a "suitcase-bomb"[2] of a pamphlet—one that managed to offend innumerable individuals in just a few pages. Huxley strove to articulate what he believed was the only viable philosophy under which an organization like UNESCO could operate: what he called scientific humanism. He could not imagine that anyone would object. No extant religion or political or economic doctrine could be endorsed by an international agency like UNESCO. Instead, he affirmed,

> UNESCO must constantly be testing its policies against the touchstone of evolutionary progress. A central conflict of our time is that between nationalism and internationalism, between the concept of many national sovereignties and one world sovereignty. Here the evolutionary touchstone gives an unequivocal answer. The key to man's advance, the distinctive method which has made evolutionary progress in the human sector so much more rapid than in the biological and has given it higher and more satisfying goals, is the fact of cumulative tradition, the existence of a common pool of ideas which is self-perpetuating and itself capable of evolving. And this fact has the immediate consequence of making the type of social organization the main factor in social progress or at least its limiting framework.[3]

He rejected traditional religion and communism out of hand; he accepted evolutionary theory as an uncontroversial fact; he advocated birth control openly, as being of great value in controlling population growth, which he saw as a major threat to the quality of human life everywhere. He upset almost everyone, without meaning to, if only because he had issued his own philosophy under the guise of a UNESCO publication. He managed to pacify most of the offended by offering to insert an erratum slip in every pamphlet, stating that this was a personal, not an official, publication.

Despite this faux pas, Huxley was elected the first director-general of UNESCO in late 1946, for a two-year term. He brought the organization into being, established its general direction (and independence), and secured its multinational character. Its concerns were the promotion of peace and international cooperation to foster conservation, the protection of nature, the education and welfare of people, the exchange of ideas, and the advancement of science for the good of humanity. Indeed, he advocated the use of scientific knowledge, backed up by the authority of UNESCO, to try to eradicate the worst evils of the modern world.

> There have been all the applications of science [he wrote some years later], leading to a new and more comprehensive view of man's possible control of nature. But then there was the rediscovery of the depths and horrors of human behaviour, as revealed by Nazi extermination camps, Communist purges, Japanese treatment of captives, leading to a sobering realization that man's control over nature applies as yet only to external nature; the formidable conquest of his own nature remains to be achieved.[4]

There was no doubt in Huxley's mind that UNESCO was a powerful new weapon. The broad focus and explicitly evolutionary bent established by Huxley (despite protests to the contrary) led directly to an endeavor that had long-lasting, indelible effects on ideas about the evolution of human beings. Even though Huxley was no longer director-general at the time the project was undertaken, his influence was still considerable; the stamp of his convictions and personality comes through clearly.

In 1949, the year after Huxley's term as director-general ended, a resolution was passed in the United Nations Economic and Social Council, asking UNESCO to "consider the desirability of initiating and recommending the general adoption of a program of disseminating scientific facts designed to remove what is generally known as racial prejudice." In response, the Fourth Session of UNESCO's General Conference adopted the following resolutions for its 1950 program: "The Director-General is instructed: to study and collect scientific materials concerning questions of race; to give wide diffusion to the scientific information collected; to prepare an educational campaign based on this information."[5] What was called for was little more than a rewrite of Huxley and Haddon's antiracist book, but in more general terms that could be endorsed by anthropological and biological organizations worldwide. The idea was pure Huxley, in its high-mindedness, in its naive belief in the power of science to persuade, and in its determination to combat the racism that had led to the dreadful nightmares of the last war. The suggestion to form such a committee may not have been Huxley's; it was certainly an idea of which he approved. The task of organizing the committee to draft the UNESCO Statement on Race fell to a Brazilian anthropologist in the Social Sciences Division of UNESCO, Alfred Ramos. Ramos set up the committee, but the man who actually wrote the words of the statement, the rapporteur, was a British anatomist residing in the United States then known as Ashley Montagu. He was an ironic choice.

Montagu was a man who had come to this prominence by a tortuous route.[6] He had started life as a working-class boy, called Israel Ehrenberg, in London's East End. The forename Moses was added at one point, in an attempt to help cure him of some childhood disease. His name declared his Jewishness unequivocally, in an era in which English Jews were often discriminated against, although not actually murdered as in Germany.

A sensitive, intelligent boy, he used books as an escape from the aspects of working-class life that he found brutish and crude. He haunted the street market of Whitechapel, hungrily buying any and every book he could afford. Those on anatomy and physiology, with their elaborate plates, especially fascinated him. But his bookishness was regarded as detrimental by his father. Their conflict reached a head during Montagu's adolescence, around the time when many working-class youths leave school to take up jobs. Hoping to free the boy of this unnatural obsession, so he might do something normal like go into business, the father tore up his son's precious secondhand books. This terrible blow hardened Montagu's resolve to escape, by any means, from a life where people could behave so callously, where knowledge and learning were scorned. As soon as he could, Montagu responded by defying his father and entering University College, London. He was accepted to work toward a degree in anthropology with one of the most renowned anatomists and anthropologists of the day, Grafton Elliot Smith. It was a triumph of the self-educated man.

A year later, he took the first of several steps to repudiate his background, his father, and his father's world: he changed his name, leaving behind all formal traces of his family and his Judaism.[7] Several of his close friends adopted a similar strategy for deflecting anti-Semitism at

about the same time. Montagu selected the grand-sounding Montague
Francis Ashley-Montagu, a choice that, by virtue of its hyphenation,
had a upper-class ring to it. It did not identify its possessor's religion,
unlike Moses Israel Ehrenberg, but linked him with Lord Montagu of
Beaulieu, whom he admired as a poet and thinker. The Ashley was after Edwina Ashley, a prominent society beauty who later married the
Earl of Mountbatten. They were the sort of parents that, metaphorically, he might have chosen at the time.

Montagu was still restless: University College was not the paradise
he had supposed it would be. He was learning a great deal, from some
of the most revered scholars in England, but all was not well. He had a
severe falling-out with Elliot Smith when he challenged one of his professor's pet theories in public. In any case, his assessment of Elliot
Smith by this time was not flattering: "pink and pompous" and not
knowledgeable enough[8] to teach either physical anthropology or general human anatomy. Astonishingly, for one so driven, Montagu left the
university without completing his degree. Indeed, he left his homeland
in disgust after the failure of the General Strike in 1926, going first to
America and then Italy.

After some unsettled years, he ended up in New York in the graduate program in anthropology at Columbia University under a brilliant
man, Franz Boas. He also spent a lot of time with William King Gre-
gory, a prominent paleontologist at the American Museum of Natural
History with close ties to Columbia; his influence was so strong that
Montagu refers to it as the beginning of the "remaking of Ashley Montagu."[9] Montagu completed a thesis on the Australian Aborigines in
1937 and quickly began publishing books and articles. A few years later, in 1940, he transformed himself yet again: becoming a U.S. citizen
and changing his name for a second time, dropping Montague Francis
altogether, taking Ashley as his forename, and removing the hyphen
between Ashley and Montagu.[10]

There is no doubt that Montagu was a self-made man in a very deep
sense. Family, religion, class, mentors, country, and name were all discarded, some not once but twice. He never dropped the upper-class
British accent and manner he had cultivated, remarking recently that
he is frequently taken for a lord. And yet he seems to carry within himself a deep dissatisfaction that he projects onto others as unfair criticism, a heightened sensitivity to possible racism that drives his
professional life.

By the time the UNESCO project was being started, Montagu had
already built a reputation for himself as an anatomist interested in race.

His new role, as rapporteur for the prestigious panel, placed Montagu at the center of a hurricane of controversy that might have ruined a man with less self-confidence. It was the cornerstone of Montagu's career. Ever afterward, he has been an unquestioned authority on race and against racism. He published and spoke prolifically on the subject of race, summarizing and presenting others' works masterfully, editing and republishing his own and others' articles with passionate forewords. Among his best-known contributions to anthropology are synthetic and often insightful reviews, conducted with energy and conviction and an unerring choice of timely topic.

Montagu was selected for the UNESCO task probably because he was already known for his books and articles, both scholarly and popular, on race. He was a highly visible warrior in the war against prejudice and "racialism," as racism was often called at the time. Indeed, some of his contemporaries regarded him as self-promoting,[11] perhaps because of the publicity that seemed to attend many of his actions and writings. One of the best known of these was his 1942 book, *Man's Most Dangerous Myth: The Fallacy of Race*, in which he argued:

> In our time, the problem of *race* has assumed an alarmingly exaggerated importance. Alarming, because racial dogmas have been made the basis for an inhumanly brutal political philosophy which has already resulted in the death or social disfranchisement of millions of innocent individuals; exaggerated because when the nature of contemporary "race" theory is scientifically analyzed and understood it ceases to be of any significance for social or any other kind of action. . . . It is highly desirable, therefore, that the facts about "race" as science has come to know them should be widely disseminated and clearly understood.[12]

This was exactly the intent of the UNESCO Statement on Race, so Montagu was an obvious, if ironic, choice.

The committee was international. It included Montagu himself, a Briton transplanted to the United States; Juan Comas, the father of physical anthropology in Mexico and a UNESCO employee; Ernest Beaglehole from New Zealand, who studied race and caste in Polynesia; E. Franklin Frazier, a black American who was an authority on "the Negro" in the United States and a sociology professor at Howard University; a sociologist and antiracist, Morris Ginsberg from Britain; the famous social anthropologist Claude Levi-Strauss from France; L. A. Costa Pinto, a professor of social anthropology from Brazil; and Humayun Kabir, joint secretary of the Ministry of Education, a Hindu from India. Two additional scholars, from Poland and Sweden, were unable to attend for health reasons. Ramos, the organizer, died before the meeting was convened and his role was taken over by a colleague, Robert C. Angell, an American.

As Montagu remembers the meeting,[13] the committee was unfocused, talking endlessly in circles. He impatiently burst out with his point of view and the committee asked him to write out a draft statement. By 1 A.M., he had completed what became the working draft, which was discussed and then submitted to a broader panel: Hadley Cantril, a social psychologist at Princeton; E. C. Conklin, a zoologist at Princeton; Gunnar Dahlberg, a geneticist at the University of Uppsala; L. C. Dunn, a geneticist at Columbia; Donald Hager, an anthropologist at Princeton; Otto Klineberg, a social psychologist at Columbia; Wilbert Moore, a sociologist at Princeton; H. J. Muller, the geneticist who was then at Indiana University; Gunnar Myrdal, the economist/sociologist at the University of Stockholm; Joseph Needham, a biochemist at Cambridge; and Julian Huxley and Theodosius Dobzhansky, two biologists of the new synthesis. In all, the fields of social and physical anthropology, genetics, psychology, and sociology were represented. Nonetheless, the 1950 UNESCO Statement on Race came to be widely known as the Ashley Montagu Statement.

It was made public on July 18, 1950, and was touted as "the most authoritative statement of modern scientific doctrine on the controversial subject of race that has ever been issued."[14] *Controversial* was an accurate description—if not of the subject, at least of the statement itself.

The London *Times* printed a seven-point summary of the statement that provoked a flurry of letters, including one from the Royal Anthropological Institute, which formally objected strongly to certain sections of the statement. More influential still was the correspondence column of the British journal *Man*, the organ of the Royal Anthropological Institute. The honorary editor of *Man*, a Cambridge-trained ethnologist and expert in African art named William E. Fagg, published the statement in full and invited comment from nine prominent British physical anthropologists, none of whom were part of the illustrious panel. In fact, considering their eminence in physical anthropology and human genetics, Britons were as conspicuously underrepresented on the panel and the broader advisory group as colleagues of Montagu from Columbia and Princeton Universities were overrepresented.

The correspondence over the UNESCO statement in the annals of

Man continued, literally, for years. The letters are a delicious exercise in exception, pedantry, academic insult, and confusion of factual reality with common knowledge. As the editor, Fagg, remarked drily in the opening article on the statement, ". . . certain passages were far from commanding universal agreement."[15] And Fagg ensured that these disagreements were aired.

No one minded the UNESCO statement's assertion that "all men [*sic*] belong to the same species, *Homo sapiens*" or that there exist populations within that species which differ from each other in their gene frequencies, which might be called biological races. Nor was there notable objection to the observation that *race*, as the word is commonly used, may refer to "any group of people whom [the users] choose to describe as a race [such as] national, religious, geographic, linguistic or cultural groups" which are not, biologically, races. Only three major groups or subdivisions within the species are recognized—the Mongoloid, the Negroid, and the Caucasoid—according to the statement, with little consensus on further classification. It was also agreed that "equality as an ethical principle in no way depends upon the assertion that human beings are in fact equal in endowment. Obviously individuals in all ethnic groups vary greatly among themselves in endowment."[16]

But the statement did not stop here. It advocated dropping the term *race* altogether and replacing it with the less emotionally laden and more accurate term (for colloquial usage) *ethnic group*—a suggestion derived directly from Huxley and Haddon's earlier book that Montagu had reintroduced in his 1942 book. The statement continued in words that echoed Montagu's previous writings:

> For all practical social purposes, "race" is not so much a biological phenomenon as a social myth. . . . Biological differences between ethnic groups should be disregarded from the standpoint of social acceptance and social action. The unity of mankind is the main thing. . . . And, indeed, the whole of human history shows that a co-operative spirit is not only natural to men, but more deeply rooted than any self-seeking tendencies. . . .
>
> According to present knowledge, there is no proof that the groups of mankind differ in their innate mental characteristics, whether in respect to intelligence or temperament. The scientific evidence indicates that the range of mental capacities in all ethnic groups is much the same. . . .
>
> Lastly, biological studies lend support to the ethic of universal

brotherhood: for man is born with drives toward co-operation, and unless these drives are satisfied, men and nations fall ill.[17]

The protests started immediately after publication of the statement. "It was jealousy," Montagu says. "People were hurt at not being involved."[18]

The first salvo came from Henri Vallois, a proud physical anthropologist who was the director of the Musée de l'Homme in Paris. He had not been invited to sit on the committee that wrote the statement, even though it met in Paris; his absence from the panel was called "extraordinary" by Fagg.[19] Vallois appreciated the intent of the statement, especially after the tragic events of the past few years which showed the dangers of racism. But the attempt to deny or suppress the reality of race in favor of Montagu's view that race is a myth was ridiculous and insulting, said Vallois:

> For the existence of races within the species of Man is an incontestable biological fact. . . . Pushing to extremes its tendency to minimize the existence of race itself, the manifesto declares finally (paragraph 14): "race is less a biological phenomenon than a social myth." Such an affirmation . . . contradicts not only the facts, but also the opening paragraphs of the manifesto![20]

"Denial of facts [that races exist] is no way to prevent intertribal or interrracial strife," agreed W. C. Osman Hill,[21] a renowned primatologist and prosector at the Zoological Society of London. He also objected vigorously—and on grounds that surprised some of his contemporaries—to another section of the statement.

> That range of mental capabilities is "much the same" in all races is scarcely a scientifically accurate statement. It is at most a vague generalization. It is, however, scarcely true, for temperamental and other mental differences are well known to be correlated with physical differences. I need but mention the well-known musical attributes of the Negroids and the mathematical ability of some Indian races. Of course, if we wish to assess marks for such attainments to each race, it is conceivable that all races may end up with the same total; but there is no scientific proof that this would be the case; our knowledge is far too meagre for any such categorical statement. Even if it were true that there is "no proof that the groups of mankind differ in intelligence, temperament or oth-

er innate mental characteristics," it is certainly the case that there is no proof to the contrary.[22]

The intellectual equality or inequality of races was, of course, a highly inflammatory issue ever since intelligence testing first began, and Osman Hill was correct in questioning the scientific validity of the statement's pronouncement. But some of his other remarks were simply embarrassing. Fagg attempted to dilute Osman Hill's response by pointing out that he had read only the *Times* summary, which was "if anything, less carefully worded, especially under (4) [the section on temperament and mental abilities], than the statement."[23] Don J. Hager, a member of the panel responsible for the statement, replied more acidly, pointing out:

> The only possible sense in which these alleged correlations "are well known" is that they have played such a great part in the popular folklore and superstitution of the Western world; they have also been the chief stock-in-trade of the racist ideology and outlook. . . . Professor Hill's references to "the well-known musical attributes of the Negroids," compares very favorably with the standard folk myths about the American Negro and his so-called "inborn" traits, *e.g.* "superior sensitivity to jazz rhythms," "greater sexual capacities," "child-like nature," "less sensitivity to pain," and so on.[24]

Others focused on the very optimistic assertion at the end of the Statement that humans possess innate instincts for cooperation. With the horrors of World War II as the ineradicable backdrop to the discussion, this idealistic sentimentality seems to have provoked hoots of derision and outrage. In the words of one indignant commentator, a social anthropologist from Edinburgh, K. L. Little:

> . . . what evidence is there for any scientific belief that man is born with biological drives towards universal brotherhood and cooperation and how is it adduced? To find such proof in "the growth and organization of man's communities" is merely to revive an out-moded psychological theory of instincts. The opposite conclusion, namely, that man lacks a biological drive towards universal brotherhood, etc., would be equally plausible in the light of alternative evidence of overt human behavior.[25]

Little also characterized the idea of substituting *ethnic group* for *race* in common parlance as "almost a species of magical technique based on the idea that something awkward or troublesome can be got rid of by the mere process of calling it by another name."[26] Still others objected that the substitution might even backfire, inducing more not less confusion. Stanley Garn, a physical anthropologist then at Harvard, dubbed it "lexical surgery" and predicted that it would be ineffectual in curing the disease of "dislikes, fears, and hatreds."[27]

The problems with the statement were manifest, as no doubt the editor of *Man* had intended they should be. Fagg held Montagu personally responsible for the failures of the statement. There was no previous hard feeling between them, Montagu asserts; it was a matter of prejudice.

"I think that he was a racist," Montagu says in explanation. "If you're brought up a Jew, you know that all non-Jews are anti-Semitic." He continues, "I think it's a good working hypothesis."[28]

He believes this accounts for Fagg's behavior.

In May of 1950, Fagg was pleased to announce the decision by UNESCO to convene another panel to rewrite the statement; the inclusion of greater numbers of physical anthropologists, geneticists, and human biologists seemed to promise a more accurate and less troublesome product than the Ashley Montagu Statement.

> There is no need to recapitulate the shortcomings of that document [Fagg observed]; we may recall, however, that most of them were traceable to the manner in which the whole vast field of racial studies, physical as well as cultural, was thrown open to discussion by a small group of philosophers, historians, sociologists, and others, only two of whom had any pretensions to competence in physical anthropology. . . . There can be no possible justification for regarding the criticisms which have been made as an attack upon the U.N.E.S.C.O. campaign against racialism. On the contrary, they reflected the disappointment which was felt when the document proved to be not the effective weapon which we had looked for, but a broken reed. . . . Although the constitution and terms of reference of the original panel undoubtedly involved serious errors of judgment—notably in the failure to consult even the most obvious expert opinion—U.N.E.S.C.O. deserves great credit first for undertaking the campaign at all, and secondly for the promptness with which, after the publication provoked the

criticisms, the decision was taken to convene a new meeting along lines far more likely to lead to good results.[29]

Fagg thought Darwinian evolution had grown, not a stout tree from which to fashion a sturdy weapon against racism, but a broken reed. Even he did not see it had more dangerous and sinister attributes than simple weakness.

The second panel was chaired triumphantly by Henri Vallois. Ashley Montagu was at first omitted entirely, but when Dobzhansky insisted on his friend's presence, Montagu was included as a defender of the previous statement. The new committee was heavily weighted with western Europeans, Britons, and Americans. There were eight physical anthropologists, four geneticists, and one serological anthropologist—a specialist in the distribution of blood groups among humans who was a leader in what many hoped would prove a new, nonracist approach to human classification. There were no representatives of underdeveloped nations, no non-Caucasians, and no women. Still, consensus was elusive.

In October of 1951, discord erupted with renewed fervor. A provisional text of the new Statement on Race was pulled from *Man* "at the last moment"[30] on the request of the Mass Communications Department of UNESCO for reasons which struck Fagg, as editor, as inadequate. He reported peevishly in the pages of *Man*,

> . . . an early version of the draft Statement had already been published by Professor Ashley Montagu in the *Saturday Review of Literature* (an American paper devoted mainly to literature and humour), without the consent or consultation with U.N.E.S.C.O., and moreover in a form which it was known that members of the drafting panel regarded as very far from definitive. . . . Once again, and this time through no fault of its own, [UNESCO's] campaign against racialism has gone off at half-cock. . . .
>
> The published version is illustrated with a photograph of Professor Ashley Montagu and appended to a leading article which is mysteriously titled "The Ascent of Man (1776–1951)," although it contains no further reference to the racial problems of the New World. There is no occasion to examine here the curiously inexact account which it gives of the history of the U.N.E.S.C.O. documents and the criticisms aroused by the 1950 Statement, which are attributed to racialistic bias and ignorance.[31]

Alfred Metraux, director of the Department of Social Sciences at UNESCO, attempted to defuse the situation in a subsequent letter in the columns of *Man*. He stated that he had approved Montagu's article but was unaware of Montagu's intention to publish the draft statement, particularly when it was so far from being finalized; perhaps there had been a misunderstanding. Montagu replied warmly denying any wrong-doing or possibility of misunderstanding. He asserted that he had specifically asked Metraux if he might publish the new Statement on Race in the *Saturday Review*, which had published the first statement, and his request had been approved. Further, the Mass Communications Department of UNESCO had facilitated the entire affair, indicating their tacit agreement.

By now, the sum total of the criticism of Montagu in *Man* was so hostile that Montagu resigned from the parent organization, the Royal Anthropological Institute.[32] In the meantime, the new draft statement was circulated and amended to try to ensure the broadest possible consensus.

The second Statement on Race was published by *Man* in June 1952, with the hope that it might be endorsed or at least discussed at the upcoming Fourth International Congress of Anthropological and Ethnological Sciences at Vienna. Given the controversy surrounding the timing of the release of both the first and second statements, Metraux was careful to label the new statement "provisional," although it stood without alteration in the end.

This was a more strictly biological and factual treatise that stayed well away from more philosophical observations and is in line with present scientific views of race. It asserted the unity of mankind in a single species that nonetheless varied in physical features; when groups possessed "well-developed and primarily heritable physical differences from other groups,"[33] they might be termed races. Physical differences might arise because of inheritance—genetic factors tempered by marriage customs—or environmental influences, but only genetically controlled features are legitimate factors for the scientific classification of humans. Nationality, religion, place of origin, language, and culture are not determining factors of race, the statement said, and pure races were a figment of the imagination in the modern world.

Then came the most inflammatory issues, intelligence and temperament, and the scientists sought refuge in equivocation. The new statement observed that mental characteristics were not generally used in the classification of human races. But that point aside, the panelists were unable to resolve the age-old debate over nature versus nurture.

Studies within a single race have shown that both innate capacity and environmental opportunity determine the results of tests of intelligence and temperament, though their relative importance is disputed [said the statement].

When intelligence tests, even non-verbal, are made on a group of non-literate people, their scores are usually lower than those of more civilized people . . . . Different groups of the same race occupying similar high levels of civilization may yield considerable differences in intelligence tests. When, however, the two groups have been brought up from childhood in similar environments, the differences are usually very slight . . . . The average performance . . . and the variation round it, do not differ appreciably from one race to another.

Even those psychologists who claim to have found the greatest differences in intelligence between groups of different racial origin, and have contended that they are hereditary, always report that some members of the group of inferior performance surpass not merely the lowest ranking member of the superior group, but also the average of its members. In any case, it has never been possible to separate members of two groups on the basis of mental capacity, as they can often be separated on a basis of religion, skin colour, hair form or language. . . .

The scientific material available to us at present does not justify the conclusion that inherited genetic differences are a major factor in producing the differences between the cultures and cultural achievements of different peoples or groups.[34]

The panel could not agree about intelligence, nor could they ignore in good conscience the reports of apparently valid differences in performance on intelligence tests that were accumulating in the literature. The statement's careful wording and unwillingness to support or refute either side of the debate simply focused attention more closely than ever on the thorny problem. Darwinism bent to political ends was proving a more dangerous and intransigent creature than had been foreseen.

After all, if intelligence were inherited *even in part*, as many supposed it was, then it was to be expected that different groups would vary in intelligence even as they varied in height or skin pigmentation. Variation in traits was one of the cornerstones of Darwinian evolution. But the potential to misuse and misunderstand complex concepts about the statistical distributions of intelligence within and among human groups was threateningly dire. Few scientists were sanguine about the probability that laypeople would understand how insignificant the variations in human intelligence were or how useless such information was as a predictor of individual ability. Most people place such a premium on awareness and intellectual functioning that the stakes were higher; those with impaired mental capacities are sometimes judged to be barely human, especially by those who have had limited contact with such people. It was no accident that Hitler selected the mentally handicapped ("feebleminded") and mentally ill for systematic elimination. He sensed that empathy for these groups, as for Jews, was weaker.

In years to come, UNESCO meetings would continue to thrash around on the subject of human intelligence, unable either to grasp and uproot the increasingly poisonous weed or to accept and nurture it. It was not just a UNESCO problem; it became society's bane. One who got caught in its grasping tendrils time and again, even after trying to steer clear of the vicious controversy, was the anthropologist Carleton S. Coon.

Part V
Evolutionary Politics

Pernicious Doctrines
of Racial Inequality

CHAPTER 10

CARLETON COON was a man betrayed by history.

He was the last of a type of flamboyant anthropologist-explorer, the sort epitomized by men of the nineteenth and early twentieth centuries like Richard Burton the explorer of Africa and Arabia or T. E. Lawrence of Arabia. Coon traveled widely to destinations whose chief attractions lay in their remoteness and wildness; he was the master of several languages and dialects. And, like these earlier heroes, Coon lived among tribesmen, adopting their world as his own. During the war, he was involved in spy networks and secret missions. More than once he was forced to disguise himself to escape danger. But unlike Lawrence or Burton, who experienced awkwardness at home in England because of their ambiguous social status, Coon came from a long line of self-confident patricians. The Coons were an old New England family who never doubted for a moment their own irreproachable social and intellectual standing. With a B.A. and Ph.D. from Harvard University, *the* university to such a family, he possessed impeccable credentials on all fronts.

Coon's characteristic sureness in his own ability and rightness was part of his charm and yet, in the end, contributed to the grave difficulties in which he found himself in the 1960s. He was a large, bearish man, immensely likable, full of an outrageous joie de vivre, as quick to bluster as to laugh. He had a puckish habit of recounting scatological or ribald customs of tribal peoples he had known, in a carrying voice in very proper circumstances. "Balls!" was one of his favorite expressions of disagreement. He was outspoken, knowledgeable, and somewhat larger than life.

Coon's problem was that he lived from the boisterous, imperialist, early twentieth century on into an era when colonialism became a derogatory term and most academics became outspokenly liberal. In the 1930s, Darwinian evolution had been politicized and warped into something evil and vicious by the Nazi party; evolutionary biologists opposed to this distortion responded, not by depoliticizing Darwinism, but by bending it to the cause of antiracism. But the antiracists, too, struggled with conflicting urges. Their moral beliefs demanded that they ignore or discount particular lines of evidence out of hand; their scientific ethics required that all evidence be presented, weighed, and evaluated objectively—even unwelcome findings. In the face of the patent evil of Nazi Darwinism, some scientists raced to the other extreme.

Coon chose as his life's work a topic that would become the central issue of a bitter academic feud: the classification, measurement, and investigation of the human races. Race had been the preoccupying concern of nineteenth-century workers who had forged racial categories that were usually typological ("the Sikh" and "the Andamaner"), as if the traits of one individual expressed the essence of all, and often prejudicial. Even though everyone talked of human races, and freely identified themselves or others with some particular race, just what constituted a race and how many there were remained debatable. It was the obvious and grand problem that needed to be solved when Coon was a student.

Coon struggled to infuse this traditional focus on the races of mankind with scientific measurement, to subject the data to statistical techniques of evaluation and to interpret the results in terms of modern evolutionary biology. The races were, Coon believed, local populations who had become genetically distinctive through a process of adaptation to their local environment and, perhaps, through Sewall Wright's phenomenon, genetic drift. Coon simply wanted to understand human variability, to conquer this venerable problem with the tools of the twentieth century, and to have a rip-roaring good time along the way. He traveled to place after place, photographing, measuring, talking to people from every walk of life in every corner of the earth.

Along the way, he also examined nearly all of the fossil evidence for human evolution, of which there was by now a great deal. Whereas early evolutionists had a remarkably sparse record of human evolution, during Coon's lifetime the known fossils grew from a few specimens of a few broad groups—ancient but anatomically modern humans, more ancient and nonmodern humans (like Neandertals), and very old, rather apelike human ancestors called *Homo erectus*—to a much richer and more complex mixture. An older species of ape-man, as *Homo erectus* was informally called, had been found first in Java, during the 1890s; more specimens were discovered in China during the 1920s and 1930s, and still more would be found in Africa starting in the 1960s. These were small-brained but large-bodied and upright bipedal creatures generally believed to be directly ancestral to some archaic form of human and, ultimately, modern humans. The newest member of the human family tree was a still older and more primitive group, *Australopithecus*, an African genus including several species. The fossils belonging to our zoological family, the Hominidae, had been found in reverse order—the most recent being first, the most ancient being last—which led to repeated surprises over the unexpected primitiveness of each new find. Too, methods of chronological dating were inexact until well into the 1960s, which meant that considerable energy was expended in estimating the relative antiquity of each new type of hominid. Coon, with his preoccupation with sorting out the living races of mankind, wanted to integrate the fossil record into his grand synthesis, using fossil specimens to try to trace the beginnings of the races he could recognize today. How else could he place the modern races in their evolutionary context?

He knew well that information on racial differences had been dreadfully misused; he simply believed that his responsibility, as a scientist, was to discover and report the truth, not to second-guess how others might use it. He relished the diverse personalities and physical traits of all he met; clearly he cared genuinely and deeply about some of the tribal peoples he lived with. But he was always classifying, noting features, slotting people into pigeonholes, whether he was meeting Kurdish herders in Iraq or Celtic academics in Boston. And he had names for these pigeonholes—scientific ones, not racial epithets—and an ineradicable habit of referring casually to people as he thought of them, as members of a particular racial group or subgroup. It was a habit that became more than distinctly unfashionable and, to some, offensive.

Intentionally or unintentionally, five men led Coon into the flaming controversy that blighted the end of his career as an anthropologist. One of them was his former mentor, Earnest A. Hooton, a physical anthropologist at Harvard who represented the "measuring school" of anthropology. Hooton had an indelible influence on both Coon and the entire field of physical anthropology. He was charismatic, outspoken, openly controversial. In many ways, Coon took after Hooton's persona as if he were his son. "For myself," Hooton once wrote on the subject of expressing his views on race, "I prefer to be the target of rotten eggs, rather than to be suspected as a purveyor of that odoriferous commodity."[1]

Hooton was among the first wave of researchers who, at the turn of the century, tried to establish physical types or races mathematically by measuring thousands of individuals. Whereas nineteenth-century anthropologists focused on minutiae of cranial anatomy, Hooton (and others) added in the rest of the body, describing quantitatively what came to be called the "constitution" of an individual. Hooton and his students measured, among others, some 14,000 criminals, 6,000 visitors to the Hall of the Social Sciences in the Century of Progress Exposition in Chicago, and some 3,000 law-abiding citizens of Boston—"or at any rate persons not in jail at the time of measurement," as Hooton put it sardonically[2]—a sample that reflected Hooton's keen interest in the question of whether or not certain types (racial and physical) had greater tendencies toward criminality. Hooton openly advocated eugenic sterilizations as late as 1937, on the basis of his studies of the somatotype of criminals and noncriminals, but with an important proviso:

> It is a very remarkable fact that our imperfectly segregated and classified physical types, called by courtesy "racial" types, should exist in practically identical proportions in these three widely divergent series [of individuals in his samples], and that these types should show individually certain consistencies of a sociological nature, certain occupational and educational resemblances, whether they are drawn from the cream of the population [the Exposition sample], from the middle of the draught [the Boston sample], or from its very dregs [the criminals].
>
> Certainly these findings should not be interpreted as a substantiation of any of the ridiculous and pernicious doctrines of racial inequality which have become a menace to the peace of the world and which have brought tragedy upon millions of blameless and worthy individuals. Every one of our so-called "racial" types in these series is represented by a substantial body of convicted felons at the one end of the scale and a group of eminently respectable and intelligent citizens at the other.
>
> There is no anthropological ground whatsoever for selecting any so-called racial group, or any ethnic or national group, or any

linguistic or religious group for preferment or for condemnation. Our real purpose should be to segregate and to eliminate the unfit, worthless, degenerate and antisocial portion of each racial and ethnic strain in our population . . . by the sterilization of its insane, diseased, and criminalistic elements. The candidates for such biological extinction would not be selected on the basis of Aryan or Semitic descent, blond hair or black skin, but solely on the score of their individual physical, mental and moral bankruptcy.[3]

Lest Hooton be cast as a racist on account of his eugenic beliefs, it is important to appreciate that he issued his own statement on race in 1936, at a Conference on the Alien held in Washington, D.C. It was his summary of "the best consensus of scientific anthropological opinion upon what races are and what they connote."[4] It ran to ten pithy points, whose content closely paralleled those thrashed out more than ten years later by the UNESCO committee. Today, Montagu has no recollection of Hooton's synopsis[5] and denies that it was consulted by the UNESCO committee, so the resemblance may be purely circumstantial.

Thus, in his way, Hooton was fighting racial prejudice, too. Yet the bulk of his work was still terribly typological; indeed, his goal was to construct good, valid, mathematical "types" for the various races. Hooton believed races varied in head shape, eye, hair, and skin color, a point with which anthropologists would agree today (with the reservation that the patterns of variation are more complex than Hooton probably perceived); what would be more strongly challenged today was his belief that intelligence and emotional or psychological traits vary similarly. His oversimplistic mode of thinking about human types and variability was incubated in his students and targeted Coon, one of the most prominent, for burning accusations at the height of his career.

Another character in the unfolding drama was perhaps the most extreme typologist among those influenced by Hooton: William H. Sheldon. Sheldon was a psychologist by training; he headed up a long-term project at Columbia and Harvard, where Hooton was professor, that tried to correlate constitution (somatotype) with temperament or behavioral attributes.[6] A handsome and charming man, Sheldon managed to maintain cordial relations even with those who openly disagreed with the conclusions of his research. Thus, his personality deflected the righteous anger that might have been directed at him onto others, including Coon.

There is no particular evidence that Sheldon was explicitly racist in his beliefs, although his research offered abundant opportunities to those who were. In 1940, Sheldon published a scheme for describing the human physique that would become famous. He identified three extreme body types: *endomorphs* were round, fat, squat individuals; *mesomorphs* were square-built, muscular and athletic; and *ectomorphs* were long, lean, and thin types. Each individual to be classified was assigned a code of three digits from 1 to 7 that expressed the extent to which that individual showed endomorphy, mesomorphy, or ectomorphy, with the respective extremes being 711 (hyperendomorph), 171 (extreme mesomorph), and 117 (hyperectomorph). Sheldon believed that these three types or qualities were tripolar opposites that could be thought of as representing the three corners of a triangle. Each individual's physical type could thus be expressed by a point on that triangle placed the appropriate distance from each corner.

This system of physical stereotypes was benign. But Sheldon worked out a parallel system for categorizing mental or behavioral attributes— again based on three extremes, with a scoring from 1 to 7—that was less neutral. He called these temperamental qualities *viscerotonia* for the emotionally warm, comfort- and food-loving, passive, and complacent type; *somatotonia* for the aggressive, extroverted, dominating, and energetic type; and *cerebrotonia* for the intellectual, introverted, shy, and oversensitive type. Not surprisingly, endomorphs (overweight, round body types) tended to be viscerotonic, athletic mesomorphs tended to be somatotonic, and tall, skinny ectomorphs were likely to be cerebrotonic. The descriptions were sufficiently generalized and vague to sound plausible, in the same way that horoscopes often strike the uncritical as accurate. The problem was that Sheldon was utterly serious and believed earnestly he was uncovering evidence of important genetic relationships.

Because the results of research by Sheldon and others who used his system appear to give a scientific justification to the view that many mental and behavioral traits (such as criminality) are genetically determined, it is crucial to understand his methods. He began by constructing the somatotype system for "measuring"—or, more properly, scoring—physical types. He was convinced that there were mental or emotional qualities that correlated with the physical ones, so he devised an interesting strategy to discover them: he would score a sample of individuals for both physical and mental attributes and then look for statistical correlations.

But which were the pertinent mental traits? He first drew up a list of 650 "alleged traits of temperament," which was "sifted, condensed, and described as systematically as possible," ultimately reducing the list to 50 traits that "seemed to embrace all of the *ideas* represented in the original 650."[7] No measurement of mental or emotional attributes was employed, only a uniform scoring conducted by a single investigator, himself; he examined and photographed 33 male college students to score somatotype and conducted lengthy interviews with them on the basis of which he scored them on the 50 mental traits.

The resultant data were then analyzed statistically in a search for meaningful correlations. In fact, using statistical correlation on data that are ordinal—arbitrarily assigned, ranked categories rather than actual, continuous measurements—violates one of the primary assumptions of the technique. Such misuse tends to falsely inflate the correlation between any two scores, suggesting causal relationships where there are none. Thus the validity of his findings is open to serious question. (No one since has been able to demonstrate in any rigorous fashion that there is a strong general correlation between physical and mental or psychological traits.) At best, Sheldon demonstrated an internal consistency in his judgments.

From the temperamental scores, Sheldon constructed clusters of traits that correlated with each other and not with traits in other clusters. He found three such nuclei, which incorporated 22 of the 50 traits. After a painful series of attempts to refine and define these traits, Sheldon then searched for others that fit the emerging pattern until he had a group of 20 traits that he felt described each of the three mental extremes. Like the physical traits, these could be expressed as a triangle.

Over the course of the first five years, he and his team scored the physical and mental attributes of 200 college men, 100 delinquent, homeless, or maladjusted youths, and a series of individuals who presented themselves at area hospitals for various problems. Sheldon reported finding a statistical correlation of about 0.80 (out of a perfect score of 1) between the primary, physical component and the "apparently secondary temperamental attributes"[8] in this sample. Unaware of the gross statistical flaws in his method, he felt this and other statistical tests validated the reality of what his "constitutional psychologists" were getting at.

He respected and believed in his technique, but he was flippant about the qualifications of its practitioners. Training in somatotyping was described by Sheldon as being

about comparable to training in cattle judging. To start with, it is necessary to be interested in cattle, and observant of them. In perhaps a year's time an observant person who has a true interest in human stock, and some background in anatomy or physical anthropology, can be taught to carry out a physical analysis with dependable accuracy.[9]

To carry out the psychological evaluation, he opined, a student needed an apprenticeship with one already skilled in the technique, a rigorous background in statistics, some medical training (especially clinical work), perhaps some training in academic psychology, and the experience of psychoanalysis. "Psychoanalysis," observed Sheldon, "is a good experience for anybody, especially if carried out in the spirit of whimsical good humor and sympathetic objectivity—as one would watch the courting of sparrows. Such an experience enriches consciousness and tends to bring tolerance and perspective."[10]

Sheldon hoped to be able to explain criminal tendencies in terms of the constitution and temperament of the individuals. Both the physical constitution and the emotional makeup of an individual were, Sheldon felt, an expression of the underlying genotype. As Sheldon put it: "The somatotype attempts to supply an identification to the sum of genetic factors, which is to say, the sum of continuing influences relating to the structure of the organism. The somatotype is at worst a guess and at best a well-proved formula intended to designate a fixed genetic influence."[11] That there was always some lingering veracity in Sheldon's stereotypes lent terrible plausibility to his system. Yet it was little more than well-intentioned but ill-conceived chicanery, supported by dubious statistics and veiled with the colors of the new evolutionary synthesis. To the antiracist liberals, Sheldon's work yielded unwelcome and ill-supported conclusions.

Two of those liberals had been involved in the fiascos associated with the UNESCO statements on race. Both Ashley Montagu and Theodosius Dobzhansky had adopted the strategy of using modern evolutionary biology as a means of combating racism and shattering stereotypes. Montagu was a public authority on the topic. Fueled by the need to support his growing family (Montagu's own explanation) or by a drive toward self-aggrandizement (his critics' charge), Montagu lectured in public, on the radio, and on television, and published at a dizzying rate. He pronounced on race, on crime, on the biosocial nature of man, on the concept of the primitive, on cooperation and competition, on aggression, on human relations, on human heredity, on race relations, on the superiority of women, on intelligence, on human nature, on prenatal influences—in fact, on almost anything and everything to do with humankind. Between 1950 and 1970, he published at least nineteen books and numerous articles. All were elegantly written and uniformly entertaining, but many were repetitive. Some of his more felicitous words and phrases seem recycled from his writings into his conversation. Many publications were authoritative and perceptive reviews organized around a worthy theme; all promoted his fervently antiracist, humanitarian, and egalitarian point of view.

He and Coon both possessed substantial egos, although their styles differed radically. Coon was flamboyant and utterly confident in his impeccable New England, Protestant heritage, while Montagu was the former working-class Jewish boy who had scrambled out of his background into an upper-class persona. Changing his accent and his name were all part of Montagu's self-transformation; so, perhaps, was leaving England for the United States. Still, the two were great friends until 1945, when Montagu wrote an article critical of Coon, which Coon regarded as backbiting. They argued at their next chance meeting on the street; Coon never forgave Montagu, whom he regarded as sneaky and underhanded ever afterward.[12]

"Of course, he knew I had changed my name," Montagu says now of their many disagreements, "and this was always regarded [by him] as some sort of crime: how dare you pass yourself off as a WASP?"[13] To Montagu, Coon's response was anti-Semitic. To Coon, denying one's background as Montagu had done was unthinkable, a terrible deception. Coon's view was shown by an incident in which a Jewish girl sent him a photograph of herself and asked him, as a specialist in race, whether or not she should have her "Jewish" nose "fixed." His reply was that her ethnic ancestry and heritage were something to be displayed proudly rather than concealed.[14] But, of course, Coon had not experienced the class prejudice and anti-Semitism that had dogged Montagu's youth in England in the first three decades of the century. It is little wonder that the two came to verbal blows.

A friend of Montagu's, with similarly liberal views, was Theodosius Dobzhansky, who also played a role in Coon's professional demise. Although tactful and unaggressive over personal matters, Dobzhansky was uncompromising about science. "In science," recalls Montagu, "he didn't think that one should be diplomatic; the truth was the truth and there are ways of calling a shovel a spade or a bloody shovel. As a scientist, your job is to leave out the adjectives and get down to reality, insofar as it's ever possible to do so."[15] Dobzhansky was not one to allow for alternative viewpoints on matters of science. Having seen hardship and repression in his native Russia, he was also intolerant of views he judged politically incorrect.

Starting in the early 1950s, this fruit fly geneticist began to dabble in human evolutionary studies, using his position as an authority on modern evolutionary biology to pronounce on the classification of fossil humans—a field remote from his own expertise by his own admission.[16] Many physical anthropologists, abashed by their own naivete about the new synthesis, solicited advice from Dobzhansky, Ernst Mayr, and G. G. Simpson, in particular. This triumvirate did much to revolutionize that field by persuading anthropologists that it was essential to abandon typological thinking. The catchwords of the day were "population" and "variability." Evolution proceeded statistically, by altering the frequencies of different genotypes within loose coalitions of interbreeding individuals that could be called populations, but the mean or the average genotype for that population was a frankly inadequate description of the true and important variability of that group. Dobzhansky was firmly committed to the idea of the unity of mankind and expressed it with an earnest fervor.

A Russian emigré, Dobzhansky was short on formal education—he never received his bachelor's degree and did not pursue any further degrees[17]—but long on brilliance. When he arrived in New York in 1927 to work with Thomas Hunt Morgan's famous group researching *Drosophila* genetics, he had a lot to learn, including English; he knew a great deal about insect anatomy and natural history, but knew almost nothing of modern genetics and still less of statistics. The latter was a surprising failing in a man who made his reputation by demonstrating the validity of theoretical models of population change in real populations. But during his long and fruitful collaboration with Sewall Wright, Dobzhansky simply skipped the mathematical portions of Wright's work.

> I am not a mathematician at all [Dobzhansky confessed when asked about it]. My way of reading Sewall Wright's papers, which I still think is perfectly defensible, is to examine the biological assumptions the man is making, and to read the conclusions he arrives at, and hope to goodness that what comes in between is correct. "Papa knows best" is a reasonable assumption, because if the mathematics were incorrect, some mathematician would have found it out.[18]

This willingness to accept another's work on faith freed Dobzhansky to look at the big picture of evolution, to stand back and try to perceive the grand pattern. He has been described as "a 'romantic,' with a penchant for generalizing from a particular set of data to larger scientific questions, and from there on to even larger philosophical and cultural questions."[19] It was a personality trait that made Dobzhansky perfectly adapted to tackling the formidable problem of the evolution of human races, and he did, even when it also meant delivering a harsh blow to Carl Coon.

The final player in the drama was Sherwood L. Washburn, a friend and former colleague of both Montagu and Dobzhansky. Washburn was a small man, but bright, aggressive, self-assured—some have said cocky. He trained in physical anthropology at Harvard with Hooton and Coon among others. Like Coon, he came of old New England stock; in Washburn's case, his father was a professor of divinity, who may have provided the model for Washburn's strong and inflexible sense of right and wrong. He was well trained, a sound anatomist, and a man with a new vision for his field. Sheldon, whose work was fiercely criticized by Washburn, once alluded to these personality traits by describing Washburn as "a tremendous fire in the Department of Anthropology at the University of California at Berkeley."[20]

Washburn was determined to bring evolutionary thinking and functional anatomy—the anatomy of bodily systems with jobs to do—into physical anthropology. In 1951, he would christen this approach the "New Physical Anthropology," but the ideas were there much earlier. In 1948, Washburn moved from his first professional job, in the department of anatomy at Columbia's College of Physicians and Surgeons, to be a professor at the University of Chicago, an indication of his rising prominence; later still, he moved on to the University of California at Berkeley. Washburn formed a strong coalition with Dobzhansky and Montagu, both of whom had been at Columbia (Montagu as a Ph.D. candidate and Dobzhansky as a professor in the department of zoology) while Washburn was on the faculty. Washburn had almost certainly met Sheldon in the last years before each left Harvard; Washburn even posed in his underwear for somatotype photos. At Columbia, Sheldon was director of the College of Physicians and Surgeons and thus Washburn's superior at the time. Washburn eventually came to feel that Sheldon exemplified the worst tendencies of typologists like Hooton, under whose training Washburn had chafed at Harvard. Washburn was bitterly impatient with the "old guard" and their old-fashioned approaches and had deep, lingering bad feelings about the Harvard anthropologists. His criticism of Hooton was so outspoken that some regarded it as Oedipal; Hooton, deeply hurt, responded shortly before his death in 1954 by making it known that he did not want Washburn to be his successor.[21] Whether in response to Hooton's request or for other reasons, the ambitious Washburn was never offered this prestigious post or any other at Harvard.

By 1950, the elements of the tragedy were in place and the actors began to play out their predestined roles, leading to events that would scar the rest of Coon's life.

Washburn and Dobzhansky, in an evolutionary coalition, together organized the fifteenth Symposium on Quantitative Biology. Ironically, the venue was the Cold Spring Harbor Biological Laboratory in New York—of which Charles Davenport had once been the director and where he founded the Eugenics Record Office in 1904. (The Eugenics Record Office had finally been shut down in 1940 as an embarrassment.[22]) They invited an august assemblage of scientists drawn primarily from the fields of genetics and anthropology to discuss the Origin and Evolution of Man. Hooton, Coon, Montagu, and Sheldon were all there as well as many others.

The preface to the conference volume makes Washburn's and Dobzhansky's agenda explicit:

> For nearly two centuries anthropology and biology have developed almost independently, although both have been profoundly influenced by such fundamental discoveries as Darwin's theory of evolution and his finding that man is a part of nature. In our century, the development of genetics, which studies the phenomena of heredity and variation, has caused a gradual drawing together of biological and anthropological research.
>
> Man, like any other living organism, is a product of his heredity and his environment. Neither can be ignored, if we wish to reach a coherent understanding either of an individual human being or of a group of human beings such as a population or race. Nevertheless, until recently there has been relatively little contact and collaboration between anthropologists and geneticists or other biologists. The chief aim of the fifteenth Symposium on Quantitative Biology was to help establish such collaboration.[23]

In effect, it was the Princeton conference, which in 1947 had brought about the new synthesis in evolution, revisited on anthropological turf.

Interspersed with dry, data-filled papers reporting the incidence of various anatomical features among modern human groups (for example, schizophrenia in north Swedish populations, hereditary diseases in Denmark, venous patterns on the chest among Navaho Indians, frequencies of blood groups in Mediterranean peoples) were a few of far-reaching import. For many, it was an occasion to nail their evolutionary colors to the mast in public.

Washburn presented one of the first exercises in his new analytical approach to human evolution, evaluating the anatomy of living or fossil primates and humans in terms of function and evolution.[24] It was yet another blow aimed at destroying the old, typological approach and establishing in its place a "New Comparative Anatomy" analogous to the "New Systematics" or the "Modern Synthesis" in evolution. It was the wave of the future, the only future possible for physical anthropology in Washburn's view.

Of course, the American prophets of the new synthesis—Dobzhansky, G. G. Simpson, and Ernst Mayr—were there. Mayr repeated a suggestion he (and Simpson) had made less formally a few years before that was deeply symbolic.[25] He observed that the record of fossil humans was now hopelessly cluttered up and confused by the proliferation of taxonomic names; each scrap of fossil from each new site seemed (to its discoverer) to warrant not only a new specific but also a new generic name. Though Mayr had not examined any of the fossils himself, he ventured to suggest a sweeping revision: everything from the earliest ape-man to the latest modern man ought to be included in the genus *Homo*—possibly even within *Homo sapiens*. He conceded that there might need to be three species within the genus—*transvaalensis* for the australopithecine ape-men, *erectus* for *Homo erectus*, and *sapiens* for living humans and Neandertals.

The political and biological message was clearly focused on the unity of mankind throughout evolutionary history. If such diverse creatures as australopithecines, *erectuses,* and modern humans could be incorporated into one genus or species—despite brain sizes that ranged from about 400 to more than 1200 cc and equally marked disparities in body size and proportions—then the differences among the modern races of mankind, which were practically indistinguishable skeletally, were trivial.

In the discussion that followed, some of the anthropologists responded territorially, questioning (heatedly or coolly) the wisdom of lumping all fossil humans together. And too few forms can be as difficult as too many; "convenience [in nomenclature] can be pushed to such a point that it becomes a confounded nuisance."[26] Montagu wanted to form a committee to revise the classification of fossil humans. Washburn deflected some of the intensity by proposing a face-saving compromise, in which nomenclature reflected differences in adaptation. Then apes and humans would be put in separate families, because of their differences in locomotion, and small-brained (*Australopithecus*) and large-brained (*Homo*) humans should be placed in different genera within the human family. Disputes aside, the aim of the conference was beginning to be reached; the participants were discussing humans in terms of evolution and adaptation.

Coon spoke in a session entitled "Race Concept and Human Races." He and two other participants in the conference, Stanley Garn and Joseph Birdsell, had recently published a small book, *Races*, in which they hypothesized about the adaptive significance of some of the physical traits that characterize races, such as the enhanced protection against the tropical sun offered by more highly pigmented skin. It was one of the first, brilliant attempts to take an ecological and evolutionary view of modern humans. Now Coon wanted to go a step farther and deal with populations on an even more human level. Drawing on his tremendous breadth of knowledge of nonindustrial peoples around the world, Coon chose to discuss how culture may have influenced genetic changes in human populations.

The paper is interesting, amusing, and well worded; treating a mildly titillating subject like the selection of mates among humans, replete with exotic examples, brought out the best in Coon. He breezed through such venerable anthropological topics as endogamy and exogamy in marriage customs. ("The choice of spouses in any given social system follows one universal principle: the ideal or preferred mating is one which under normal circumstances will produce the minimum of disturbance to all persons in any way concerned."[27]) He described love and marriage among Arab camel breeders, various practices among people in caste systems from India to Samoa and among inhabitants of Turkish harems, bride choice in the Ona of Tierra del Fuego, and many other things. He explored the principles of "extracurricular" matings and the circumstances under which this is permissible. He also considered fertility and how often mating produces offspring; is the familiar pattern of males going off on hunts and then returning at intervals actually a means of enhancing fertility? And to what extent does nutrition, disease, genital mutilation, fatigue, fear, or even the practice of taking hot baths diminish fertility, he wondered. Finally, he passed quickly through factors affecting the survival of children, the reproductive longevity of individuals, and the question of promoting the survival of one's group at the expense of one's own survival. It was all anecdotal, it was classically anthropological, and yet it was brand-new: ethnography viewed from the perspective of evolution.

The printed discussion suggests that few speakers appreciated Coon's novel intent.[28] None of the evolutionary experts commented at all. There were only some rather testy questions about points of fact and Coon replied tartly by referring the speakers to the references cited (including their own). Perhaps to most of the audience, Coon seemed simply to be telling stories to remind his audience of the wonderful diversity of human behaviors.

In contrast, the paper presented by young Joseph Birdsell on genetic markers among tribes of Australian Aborigines, complete with mathematical models of gene flow, was greeted with excitement. *Here* was the sort of anthropology that seemed modern, statistical, population-based. Unfortunately, it was all rather abstract and altogether too tidy, as mathematical modeling is wont to be and human beings never are. Birdsell omitted any mention of all of those messily behavioral, awkwardly esoteric customs of marriage and mating that old-style, holistic anthropologists like Coon savored.

Montagu made his evolutionary and egalitarian bent clear in a long, philosophical paper about the concept of race from the modern evolutionary point of view.[29] There was little new or original in his paper. He summarized or restated the definitions of concepts like natural selection, mutation, genetic isolation, genetic drift and so on, quoting Dobzhansky, Mayr, Simpson, and other new synthesists. He criticized Coon, Garn, and Birdsell for relying on phenotypic (grossly visible) traits in their classification of human races and suggesting that, secondarily, genetic differences might be added in where those are known. Given the immature state of human genetics—Dobzhansky would list in his paper a grand total of twenty-two traits, three based on blood groupings, that were well or approximately established—Coon and his colleagues were simply being pragmatic; but then, as much later, Montagu and his confreres felt that the emphasis must be laid on the underlying genetic component and not on its expression. Genes were the stuff of evolution and inheritance, and shifts in gene frequencies provided the motor. But the point that drew the most comment was his repetition of the old suggestion that *ethnic group* replace *race* in common usage, which was persistently unpopular.

Before Dobzhansky spoke came two other extraordinary presentations. Carl Seltzer, a Harvard anthropologist, presented an analytical comparison of the somatotypes of 500 male juvenile delinquents with those of a group of nondelinquents matched for age, socioeconomic background, IQ, and "national origins," which probably meant race. His was a purely Sheldonian exercise. The other paper was by Sheldon himself.

Seltzer spelled out the logic of his inquiries clearly.

> The role of physique as a prime manifestation of the "constitution" of the human organism is axiomatic. Not only does it provide the basic framework through which the individual functions, but its importance is more far-reaching in that the individual's physique is, in the main, biologically inherited, genetically determined. Body-build being thus "a product of influences emanating from the germ plasm" is therefore a constitutional determinant in personality formation.[30]

The results of the study were clear. The delinquents were physically normal—contrary to some previous theories—but tended to be "absolutely and relatively more mesomorphic and decidedly less ectomorphic than non-delinquents. . . . In comparison with the non-delinquents, then, the delinquent juveniles should be distinguished by more assertiveness, competitive aggressiveness, forcefulness, psychological callousness (insensitive to inhibitory suggestion), adventurousness and extroversion."[31] This constellation of personality traits (not demonstrated but supposed to exist because Sheldon had found that they correlated with the physical types) seemed to explain why these individuals engaged in more antisocial acts. Seltzer concluded that there was an important biological component to delinquency.

Aware of the potential danger of such a statement, Seltzer cautioned that this finding did not mean that there was inherent criminality in the individuals or that there were fixed criminal types. It did mean, however, that the delinquents tended to have complexes of normal personality traits which "make them more readily activated towards the commission of anti-social acts. These personality traits or their combinations are by no means the exclusive property of delinquents, but they are to be found *in greater frequency* in the delinquent population than in the non-delinquents."[32] This was fodder delicious to the most prejudiced eugenicist, but for one easily overlooked factor. The key here was that "population," to Seltzer, did not have the same meaning as "population" to the geneticists and anthropologists at the conference. Seltzer's population was not a breeding group or a tribe but a study sample selected from homes for juvenile delinquents. There was no apparent evolutionary validity to his "population."

Sheldon rose after Seltzer and described briefly the rationale of his techniques and the objectives of the long-term projects of the Constitution Laboratory at Columbia, which he led. Although his influential project paid lip service to evolutionary theory, their concern was with the readily observable:

> Behind the successive phenotype presentations we assume that a genotype pattern is at work, but in constitutional work our concern has not been to try to describe the genotype as such, or to attempt to isolate specific genes. . . . Perhaps a simpler way to put the matter would be to say that the aim of the constitution project has been merely to follow the human phenotype (systematically or classificationally) in both its biological and its social relationships. The watchword at the Constitution Laboratory is this: To re-establish the broken continuity between biological and social science.[33]

In other words, Seltzer's conclusions were almost preordained. Behavior was influenced by genetic makeup and it was the task of Sheldon and his followers to find ways to explore these underlying links.

Sheldon presented no data, no results, no research at all. The tone of his paper suggests he was defensive, with good reason. Pointed questions were raised in the discussion, especially by Washburn: Does somatotype stay constant throughout life? If not, what is the validity of Sheldon's work? Isn't somatotype amenable to change with hormonal treatment, exercise, or diet? Why do male and female somatotypes show entirely different distributions on the triangular graph when they come from a single genetic population? The criticisms were devastating, Sheldon's responses unconvincing to most.

The lengthy discussion following Sheldon's presentation foreshadowed things to come. The lines were drawn then, in 1950, on what would become the pivotal issues of the battle in 1962, when Sheldon would no longer play a crucial role. Montagu criticized Sheldon sharply for wasting his time on appearances when it was genetics that mattered. He coyly veiled his attack with a subtle remark: "I want to make it quite clear that I am making these remarks, in Dr. Sheldon's terms, as a '623,' I am being 'compassionately empathetic.' "[34]

Sheldon replied in kind, observing acutely:

> Dr. Montagu is committed to a religion of environmental determinism in human affairs. This is a romantic and in some respects a courageous way of viewing the human scene. But it is dangerous, for it rests too much weight on one side of a complex balance. Because of this present fashion in some academic circles to overload the environmental basket with eggs, I think there is some need for a counter-emphasis [on the genetic component of human behavior].

Sheldon then delineated the issues, bringing the true point of disagreement into the open.

> Dr. Montagu has indicated, in another place, that he thinks the constitutional emphasis is really a veiled form of racism, or fascism. My view is that the only adequate defense against this very danger lies in looking *at* (not away from) the presenting peculiarities and attributes of all sorts and breeds of human beings. The constitutional procedures amount essentially to an organizing and systematizing of the field of individual differences. To impute to the extension of this kind of knowledge the danger of its subversive misuse by future malevolent fanatics, is perhaps to question the wisdom of looking at reality at all.[35]

And this was the heart of the debate: was it better to examine and acknowledge differences among races or to deny them? Was it as wrong to distort or hide facts in the service of antiracism as it was to subvert them to racist beliefs?

Coon and Sheldon would probably have answered these questions affirmatively; they had no fear of the facts that scientific scrutiny would reveal. The truth must be sought at any cost. But for Washburn, Dobzhansky, and Montagu, there were issues of racial differences that were better left unexplored, because such data might be misused by those who believed that racial groups could be defined in a consistent manner. That human races could *not* be neatly defined was axiomatic to these three—and, indeed, all attempts to concoct simple and scientifically accurate definitions of human racial groups had failed. Thus, trying to document racial differences was a waste of time and ran the risk of providing ammunition to potential racists. Some scientific studies were simply too dangerous, socially. It was an attitude with which Virchow would have sympathized, so many years before.

No better statement of the primary message of the conference was made by any speaker than by Dobzhansky in the concluding session, whose talk was entitled "Perspectives of Future Research." Here was a chance for one of the leaders of the new evolution, one of the great synthesizers, to point the way. Dobzhansky responded, perhaps predictably, by calling for more studies, more data on the variability of genetic factors in humans (but not of human races), more censuses of populations (which were not races), more investigation of the underlying genetic causes of observable phenotypic differences (which were not useful in classifying humans into races).

But the main point of his speech was a throwaway line, a seemingly casual observation: "Darwin proved that man is a part of Nature. But man lives in a cultural environment created by himself. This duality of man was, in fact, stated in Aristotle's famous dictum that *man is a political animal.*"[36]

After all, the objectives of the conference were political. It was not really a question of promulgating neo-Darwinism, the evolutionary theory of the twentieth century that brimmed with genetics. The meeting was an attempt to consolidate and institutionalize an approach to human evolution that owed more to political conviction than to science.

T WELVE YEARS PASSED. The United States entered a new and wrenching era. Just a few years after the Cold Spring Harbor symposium, the fuse had been lit on the explosive question of racial inequality in the United States by a quiet and tired black woman. Rosa Parks refused to move to a seat at the back of the bus in the "Negro section." This small decision, made by an ordinary woman with no history of activism, was the trigger for change.

Soon the Reverend Martin Luther King, Jr., began constructing his massive nonviolent resistance movement, with the founding of the Southern Christian Leadership Conference. And in 1960, John F. Kennedy was elected president, in part because of his support of King and the burgeoning civil rights movement. By 1962, sit-ins, civil rights demonstrations, and Freedom Riders were testing the segregation laws all over the American south. It was a deeply disturbing and troubling time for both the liberals, ardently in favor of equal rights for people of color, and the conservatives, who could not understand why customs and attitudes of a hundred years had to change. Race was the tension that was testing the fabric of American unity and democracy. It remained to be seen if the outcome would be the destruction of a once-great nation or a new strength.

And race was, as always, the concern of the anthropological community, who now felt a new relevance in their studies. On the one hand, it was seen as crucial for anthropology to be modern and scientific. On the other, it became important to ensure that anthropology was not harnessed to the cause of discrimination, as it had been in Nazi Germany. This time, anthropology would be mustered on the side of racial equality and freedom.

Simultaneously, anthropology was undergoing its own radical reorientation in theory, thanks to Washburn and his New Physical Anthropology. The new synthesis—no longer new and no longer controversial—was incorporated enthusiastically into the operating paradigm of mainstream physical anthropology. The term "biological anthropology" came into usage in some circles as a way of emphasizing the new and stronger links between physical anthropology and evolutionary biology.

The Wenner-Gren Foundation for Anthropological Research, founded in 1941, acted as a catalyst in the reorientation of physical anthropology. It was a small foundation whose funding objectives were strongly influenced by the acute personal assessments of its directors of research. The first was Paul Fejos, a flamboyant Hungarian who specialized in ethnographic films. Fejos ran the foundation between 1941 and his death in 1963; the subsequent director was his young widow, Lita Binns (later Lita Osmundsen), who remained director of research until 1986. Both Fejos and Osmundsen prided themselves on their ability to spot innovative thinkers with good new ideas, and they used their position to nurture such people.

The foundation, through Fejos and Osmundsen, encouraged Washburn's new direction by sponsoring his research (and his students') and by funding several conferences at which Washburn's ideas were featured prominently. In that era, being invited to a Wenner-Gren conference was a sign of having "arrived" in anthropology. Washburn recalled that his ideas owed much of their development to "discussions which have been held at the Wenner-Gren summer seminars for physical anthropologists, and those reading only current American physical anthropology would get little idea of the size or importance of these changes."[1] These seminars were Washburn's pet project, financed by the foundation. They were six-week sessions held for graduate students and leading physical anthropologists between 1946 and 1953. Participants discussed data, theory, and methods with keen excitement. Most of the seminars were held at Columbia University, though the last was at the Smithsonian. These were a perfect, informal setting for fomenting a revolution in thinking. Washburn's emphases on the understanding of the process of evolution and adaptation, and on testing hypotheses against hard data, infiltrated the minds of the upcoming generation of physical anthropologists.

In 1952, the Wenner-Gren Foundation began funding prestigious international symposia. These were convened at various locations until the foundation acquired the elegant Burg-Wartenstein castle in Austria, where all conferences were held from 1958 to 1981 (when the castle was sold). Food and drink were lavish, the surroundings were beautiful, and the discussions continued day after day, in both formal and informal contexts. These were intense, week-long meetings; spouses and friends were excluded. The isolation of the castle meant that there was no way for participants to wander off, following their own agenda and skipping the sessions. Burg-Wartenstein conferences proved immensely influential in anthropology. Reflecting on the period, Emöke Szathmary, a biological anthropologist, recalls:

> At minimum, the international symposia have produced "state of the art" perspectives, and, at maximum, they have resulted in major advances in understanding of the topics confronted. Most of the symposia have yielded books, thereby ensuring that the ideas generated reached a large audience.
>
> Washburn's articulation of the new physical anthropology was written specifically for the first international symposium, and his ideas were widely circulated in the volume arising from it, *Anthropology Today: An Encyclopedic Inventory* (A. L. Kroeber, ed., 1953). Accordingly the support provided by the Foundation comprised a "double whammy": it sponsored the Summer Seminars that were so important for the shift from the "old" to the "new" physical anthropology, and it made the news of the revolution available to thousands of readers.[2]

Washburn learned well the lesson that a good conference can accomplish much; he used them, and the resultant books, to become one of the undisputed leaders in physical anthropology. In 1959, the tangled problem of the classification and nomenclature of fossil humans arose at a meeting of the American Institute for Human Paleontology. With Wenner-Gren support, Washburn and several others wrestled with the issue at the foundation in New York and again at the meetings of the American Anthropological Association (AAA) in 1961, where he was elected president of the association. Having clarified matters in his own mind, Washburn organized an international Wenner-Gren conference at the castle for the summer of 1962: "Classification and Human Evolution." Coon was not invited despite his lifelong concern with classifying humans, perhaps because Washburn felt he had nothing new to contribute.

The conflict was about to erupt into the open. Coon was already being pushed aside by Washburn. And when Coon published his magnum opus, *The Origin of Races*, in October of 1962, Washburn would express his abhorrence of Coon's conclusions in his inaugural address as president of the AAA on November 16, 1962. Predictably, Montagu would side with Washburn, against Coon. During and after his stint on the UNESCO panels, Montagu had been chairman of the department of anthropology at Rutgers University. After stepping down, he continued to publish strongly worded, openly antiracist books and articles for popular and professional journals, and in 1962 was a Regents Professor at the University of California at Santa Barbara. Montagu was a well-known and outspoken figure, one who had often appeared before the public as a political warrior on matters or race and human nature.

Theodosius Dobzhansky, too, was Washburn's ally. Dobzhansky enjoyed tremendous esteem as a leading geneticist; he was a silver-haired, strong-willed personality whose participation in any conference on genetics was de rigueur. By odd coincidence, in 1962, he was elected president of another professional association, the American Association for the Advancement of Science (AAAS). Ever since the Cold Spring Harbor conference, Dobzhansky had been routinely consulted about matters of human evolution and classification, partly perhaps because of his friendship with Washburn and partly because the problems captured his imagination. In April of 1962, he published a classic book placing the fossil record of human evolution in an evolutionary context—a book so well received, well written, and respected that it was required reading for students of anthropology for at least the next fifteen years. It was called *Mankind Evolving*.

In it, Dobzhansky took a clear line about race among humans that has stood up well to the test of time. Races—genetically and therefore physically populations of humans—exist: "Race *differences* are facts of nature which can, given sufficient study, be ascertained objectively."[3] In Dobzhansky's eyes, races constitute subspecies of *Homo sapiens*. Because the races are fully interfertile, because different races frequently live together in the same regions, and because humans are so mobile, pure races do not exist and the races cannot diverge into separate species. Admixture is a continuous process. In short, the human species is *polytypic*: that is, a species with several different phenotypes or forms. "Civilization causes race convergence, due to gene exchange, to outrun race divergence," Dobzhansky noted.[4] In all probability, some of the characteristic features of human races are adaptations to different climatic regimes and environments; others are more probably caused by genetic drift and chance differences in reproductive rates among what were once small populations. But his take-home message about race is this:

> That people may be equal without being alike has been repeatedly stressed in this book. Equality is a precept, similarity or dissimilarity a percept. Strictly speaking, science does not tell us whether people should or should not be equal, but it does show what consequences result from equality or inequality of opportunity, given the human diversity observed. . . . Denial of equality of opportunity stultifies the genetic diversity with which mankind became equipped in the course of its evolutionary development. Inequality conceals and stifles some people's abilities and disguises the lack of abilities in others. Conversely, equality permits . . . an optimal utilization of the wealth of the gene pool of the human species. . . .
>
> Race bigots contend that the cultural achievements of different races being so obviously unlike, it follows that their genetic capacities for achievement must be just as different. . . .
>
> The decisive point is, however, that nobody can discover the cultural capacities of human individuals, populations, or races until they have been given something like an equality of opportunity to demonstrate these capacities.[5]

It is an admirable and hopeful statement, the validity of which cannot be challenged.

In the preface, Dobzhansky thanks Coon, among other "friends and colleagues for critical reading of my manuscript and for their suggestions, corrections, and emendations, by which I have profited enormously."[6] Coon is singled out as having helped with the chapters on race, polymorphism, class, and caste. He inscribed a copy of his book and sent it "To Dr. C. S. Coon, with warmest regards from the author."[7] On the eve of a bitter feud, there was only cordiality.

Coon's book, following so close on the heels of Dobzhansky's, brought him as much infamy as Dobzhansky's work won praise. The status of the participants in the fray was lofty; their aims, admirable; their clash, tragic.

Coon, too, had maintained his renowned position in his field during these years, while he tried to complete a major segment of his life's work. It was to be a massive work on the ancestry of the living races: the last word on the evolution of modern humans, one that would incorporate all of the fossil evidence from around the world. To that end, Coon traveled widely, visiting fossil sites, examining original specimens, and talking animatedly to the discoverers. His book would place the attributes of the modern races in an evolutionary and adaptive context, following up on some of the ideas in the small book he had written years before with Stanley Garn and Joseph Birdsell. He would at last integrate the ideas of the new synthesis with the data on human race.

Coon's relations with Washburn had soured during those years. According to Coon's autobiography, things went wrong after Hooton's death in 1954. Hooton, the grand old man who had raised them both intellectually, was dead; now Coon was metaphorically the elder brother, who (in Washburn's eyes) was the heir to the mantle of racial typology that Hooton had worn. Coon thus became the target for increasingly acrid attacks. Washburn seemed to equate typological thinking with racism and attacked both with an almost religious fervor.

Lita Osmundsen, who assumed directorship of the Wenner-Gren Foundation in 1963, recalls that there was still a personal respect between the two men. But Washburn was so avidly antiracist that he could not tolerate any whiff of racism in another's work, she says. Too, Osmundsen thinks that Washburn transferred his intense disapproval and dislike of Sheldon's work onto Coon, as if the fact that they shared Hooton's influence and a Harvard background contaminated them both equally: "Sheldon was the real target," she believes.[8] Oddly enough Dobzhansky was less denigrating of Sheldon's work, conceding "it still looks as if there are important biological constants underlying the observed variations of the human physique and temperament. The urgent task in this field is to discover methods of detection and quantification of these constants."[9]

By 1962, Sheldon had long since left Columbia for the University of Oregon Medical School, closing up the Constitution Laboratory at Columbia in 1959. He continued follow-up studies on some of the more than 46,000 individuals he had typed, however, and continued to develop his theories on the genetic and constitutional basis of temperament. Although Sheldon's work met with considerable criticism, it remained a major (if controversial) theory of personality that was presented in psychology textbooks for years to come.[10] But he was less visible in anthropological circles than formerly and his approach attracted few followers among the younger generation of anthropologists. If he was the target, as Osmundsen believes, he was nevertheless spared the brunt of Washburn's furious disapproval and Dobzhansky's and Montagu's relentless accusations.

Washburn's distaste for Coon's (or Sheldon's) research may have been profound but Coon was still highly regarded and was by now a full professor at the University of Pennsylvania. Coon was invited to several of the elite Wenner-Gren conferences (but not the one organized by Washburn on classification) and received two grants from the foundation for his work. In a parallel to Washburn's and Dobzhansky's success, Coon was elected president of the American Association of Physical Anthropologists (AAPA) in 1961, for a two-year term. He replaced a friend and colleague, a black anatomist from Howard University named W. Montagu Cobb. Cobb was a prominent spokesman for his community and would later become the first black president of the National Association for the Advancement of Colored People (NAACP).

Coon's serious exposure to the poisonous controversy began at the May, 1962, meeting of the AAPA, when he took the chair for the first time and found that some distinctly nonacademic topics were on the agenda.[11] His second-in-command, T. Dale Stewart, was an M.D. and Ph.D. known as the "fossil man" man at the Smithsonian. Stewart's role was important, for he was more skilled than Coon at political maneuvering and running meetings adroitly, whereas Coon's impulse was to plunge ahead, like a bull at a gate. Between them, the two managed to derail a resolution that they felt was of dubious value: this was a proposition that all races were of equal intelligence.

Coon's (and presumably Stewart's) objection to the resolution was that this was a matter of scientific fact, not of opinion; voting on it was a specious exercise, akin to voting on whether the sun would set in the west or the east. The truth was the truth and would remain so, silly resolutions notwithstanding. Unfortunately, the facts that would prove what the truth actually *was* were still lacking, as they had been when the UNESCO panel stumbled over the same point some ten years earlier.

There were (and are) various techniques for measuring intelligence, some of greater usefulness than others, but all extremely sensitive to the educational background, ethnicity, and socioeconomic class of the individual being tested.[12] The abundant evidence of the variance in absolute brain size among races was virtually useless, for there was not (and is not) any demonstrated correlation between brain size and intelligence among humans. In other words, having a larger cranial capacity inside your skull is no indicator of your innate intelligence. In fact, nearly all of the variance in brain size can be explained as a function of the differences in body size among human races. Another telling point is the cases reported from time to time in the medical literature of people of entirely normal intelligence—some successful college students, professors, or other professionals—who are discovered, perhaps during a check for injury following an accident, to have much smaller than normal amounts of brain tissue (and larger quantities of cerebrospinal fluid) inside normal-sized heads. These individuals are living demonstrations of the tremendous plasticity and redundancy of the human brain.

If pure races could be defined and found (they cannot, although Coon may have been more optimistic about this point than anthropologists today), and if intelligence can be meaningfully measured cross-culturally (which is still problematic), then theoretically data could be gathered that would demonstrate whether the races had statistically equal intelligence. For Coon and Stewart, the lack of hard evidence meant that the relative intelligence of different races was no topic for a vote. Coon remembers that former AAPA president W. Montagu Cobb—one of the two blacks present—supported him on this point and the issue died for the moment.

But race was not removed from the agenda. After the regular business meeting had adjourned and many participants had already left, some of the younger members asked for an extra session to bring up new business: "This of course was the usual trick that minorities use to get their way," Coon remembered[13]—meaning, by the word *minorities*, groups with views not favored by the majority rather than racial minorities. The resolution they hoped to push through was one condemning Carleton Putnam for his recently published racist book *Race and Reason*.

Putnam was not an anthropologist by any stretch of the imagination; he was a businessman (ex-chairman of the board of Delta Airlines) and a sometime author of the biography of Teddy Roosevelt. He was a remote relation of Coon, from another old New England family, and he had written to Coon to verify some material about races that Coon had published in preliminary form. Putnam reported the information in his book, although it was neither a major theme nor a primary justification for Putnam's views. Those views, and the book as a whole, were clearly racist. Putnam was convinced that blacks were genetically inferior to whites and should be separated from whites if not actually returned to the African continent. Perpetuation and even strengthening of the segregation laws, which existed but were being challenged daily, was the only hope for mankind, Putnam argued.

It was an offensive and misguided book that interpreted accurately reported data in a biased way. Coon had read it carefully, and had found "nothing actionable"[14] in it, but he suspected that few attempting to rail-road the resolution through had been so meticulous. He decided to confront the problem head on, asking how many of those present had actually read Putnam's book and not just Putnam's inflammatory pamphlet? Only one man raised his hand. It seemed that a lot of hotheaded talk and facile condemnation, based on hearsay and rumor alone, were at work.

Coon chastised the group for scientific irresponsibility; they had permitted themselves to be brainwashed into condemning a book's contents without having had the courage to verify their facts. It was the worst sort of mind-rot Coon could evision. It was *1984* come to life, and Coon wanted nothing to do with anthropologists "apparently as brainwashed as Pavlov's puppies."[15] Furious, he offered his immediate resignation, stomping off and leaving the meeting to discuss and pass any "craven"[16] resolutions they liked, but not under his leadership. He was even more irate when he found, some days later, that the resolution against Putnam had been passed and that Coon's resignation had been refused—meaning that the condemnation of Putnam appeared in print under Coon's signature.

Mentally and physically weary with such political struggles, Coon resigned from his position at the University of Pennsylvania to take an early retirement. Rumors flew that he had been fired for his views on race—he was contacted by Kermit Roosevelt, Jr., among others asking if this were true—but Coon denied them. His formal retirement was purely voluntary.

But Putnam would come back to haunt Coon again. Part of the problem was the similarity of their names and their familial relationship, which was still mentioned thirty years later by various scholars and was taken by some as evidence of collusion.[17] That Coon's well-known brother, Maurice, had Putnam as a middle name suggested a closeness between the two Carletons that seems to have been unjustified. And some apparently suspected that Carleton Putnam was either a pseudonym for Carleton Coon or his mouthpiece, although no evidence has ever been produced.

Coon soon found himself standing against the wall as the firing squad took aim. Applying evolutionary theory to human races, as best he could, had earned him a harsh sentence. He would be denounced for Putnam's words, condemned for Hooton's typological thinking, and executed for Sheldon's tacit assertions that criminal behavior had a genetic component.

To Padlock the Mind

CHAPTER 12

IF COON HOPED his resignation would defuse the situation, he was wrong. History had overtaken him.

In October of 1962, he published the book he had worked on for many years, *The Origin of Races*. Like Darwin's *Origin of Species*, which its title echoed, Coon's book was based on a lifetime of observations and cogitations and it was full to the brim with information and new synthesis. It was 792 pages of text, with 32 plates, maps, tables of data, and an extensive bibliography. The book appeared with advance praise from Julian Huxley, G. G. Simpson, and Ernst Mayr—three of the architects of the new synthesis—as well as from important physical anthropologists like William Howells, William Straus, and Lawrence Angel. The editor and copy editor had struggled heroically with Coon's academic prose, rendering this momentous book intelligible to the average reader. As it turned out, Coon's prose was not only intelligible, it was incendiary.

His research plan, as explained in the book, was simple. He had examined the origin of the major modern human races by compiling all of the fossil evidence from the different geographical regions that were the original homelands of those races (Africa for Congoids or what are now called Negroids,[1] southern Africa for Capoids, Asia for Mongoloids, Australia for Australoids, and Europe for Caucasoids). During the research phase, he realized that a scientist writing in the 1940s, Franz Weidenreich, had hit upon an important theory that Coon came to believe was probably correct.

With far less fossil evidence than Coon had to rely upon, Weidenreich had nonetheless observed that the species prior to our own, *Homo erectus*, was spread around the Old World as are modern humans. More than this, he saw physical resemblances between the Chinese skulls of *Homo erectus* and those of modern Mongoloids; a similar continuity could be traced between the Javan *Homo erectus* crania and those of modern Australian Aborigines; another linked Neandertals from the Near East with modern Europeans. Seeing these continuities was sometimes difficult; demonstrating them was even harder. These linking traits were often complex matters of shape in various parts of the skull that were difficult to quantify exactly, but both Weidenreich and Coon wrote when most physical anthropologists were unsophisticated about quantitative and statistical procedures. "Evolution in man did not proceed in occasional jumps [and discontinuities] as some students want us to believe," Weidenreich noted. "Quite the contrary, the continuity of the line seems amazing when the scarcity of the fossil human material is taken into consideration."[2] What this continuity led him to envision was that *Homo erectus* was already differentiating into the regional populations that eventually became the modern races. Over time, these regional populations evolved in parallel toward the "goal"[3] of modern humans—a turn of phrase that reveals a lingering progressivist or teleological sentiment in Weidenreich's mind that today sounds false.

Of course, in his theory some interbreeding between populations was possible during this process; since all modern humans are demonstrably one species, their ancestors cannot have been otherwise. In fact, Weidenreich specifically asserts the unity of the living races of man and explicitly denies that he is suggesting that the different races had separate origins from different species of apes.[4] This is a means of distancing himself from a discredited, foolhardy, and inaccurate nineteenth-century theory known as polyphyly or polygenesis.

Weidenreich's theory had the interesting effect of placing the origin of races and racial differences far back in time. These were not recent evolutionary developments, his work implied, but deeply rooted, long-standing distinctions. But his work was too typological in orientation; he had also underestimated the variability within each regional group. His ideas had been swept away in the rejection of everything old-fashioned that accompanied the intellectual invasion of the new evolutionary synthesis.

Coon believed he could weld the two theories—Weidenreich's and neo-Darwinism—into one powerful concept. He had listened to Mayr, Simpson, Dobzhansky, and Washburn urging a new way of thinking and a new emphasis on evolution in *variable populations*, not individuals, over time. He had read Weidenreich, and followed his footsteps in reviewing the features of every scrap of fossil human known to science in the 1960s. He had shared Weidenreich's vision of evolutionary continuity within geographical regions, too. For Coon the resolution was a theory that synthesized continuity *and* change. For him, the races developed long ago, perhaps a half a million years before the coming of modern man, because of their relative isolation in different parts of the world. As each local population or race evolved in parallel toward modern humans—a phenomenon that Coon saw as primarily one of increasing brain size—they retained some regional distinctiveness while yet maintaining genetic continuity with the rest of the species.

The trick was to explain, in terms of modern genetic theory, how the regional populations could develop distinctive traits (i.e., different frequencies of suites of genes) while yet remaining part of the same species. Coon's explanation was a bit crude, but not far in essence from today's thinking. He emphasized the geographical separation and isolation of the different protoraces and their need to adapt to local climatic and environmental regimes. The different, local environmental demands—and distance—would have acted as partial barriers to interbreeding, enough to permit the evolutionary development of different frequencies of various genetic traits. This need for local adaptation was balanced by persistent, but relatively small-scale gene flow, or what Coon called "genetic contact with its sister populations,"[5] which prevented speciation from occurring and kept those evolving populations as parts of a single, polytypic species.

Coon's measure of modernity—the credential needed for inclusion in our species, *Homo sapiens*—was a brain size of about 1250 cc, like modern humans'. Fossils of large-brained specimens occurred in different regions at different times, leading Coon to believe that the five races evolved in parallel, but at varying rates, toward the threshold of modern humanity. As he explained:

> Although the component populations of a polytypic species evolve as a unit, they cannot do so simultaneously since it takes time for a [favorable] mutation to spread from one population to another. . . . Related populations, which in our case are subspecies, passed from . . . *Homo erectus*, to . . . *Homo sapiens*, at different times, and the time at which each one crossed the line depended on who got the trait first, who lived next to whom, and the rates of gene flow between neighboring populations.[6]

But who *was* first? According to the evidence available to him at the time, the oldest skull with a large braincase was Mongoloid, a *Homo erectus* from China which he called by its older taxonomic name, *Sinanthropus*. The record of fossil humans in Asia was improving rapidly as Coon worked in the late 1950s and early 1960s, and it seemed quite certain to him that the Mongoloid race of *Homo sapiens* evolved from the Asian *Homo erectus*.

> The only serious doubt that remains is this: did Sinanthropus alone and unaided undergo the mutations in the central nervous system, and probably also the endocrine system, that transformed him from *H. erectus* into *H. sapiens*, or did someone else who had earlier undergone this process assist him through [genetic] mixture?[7]

Caucasoid fossils from Europe of comparable age to the Asian *Homo erectus* were rare and fragmentary; the earliest Caucasoid cranium judged to be sapient was somewhat younger (less ancient) than those from Asia. But here Coon fell into an intellectual trap—a common one at the time—of equating stone tools with particular human groups, as if the wholesale copying and borrowing of technologies and stylistic elements were impossible. Rather than stick to the fossil evidence, Coon argued that stone tools found in China attest to an early

> extension to the east of the Caucasoid geographical range [into] the homelands of Sinanthropus and the Mongoloids. . . . According to this interpretation of the flints . . . , some Caucasoids similar to those we have seen in Europe entered central and northern China from the West, and mixed with the local population, and left their tools behind them when they died. If the Chinese population had not yet crossed the *erectus-sapiens* barrier, this injection of genes could have given them the chromosomal equipment to initiate such a transition.[8]

At the very least, Coon was here invoking technological evidence which he did not consider when writing about other geographical regions. To some, he seemed to be scrabbling to make Caucasoids the first to achieve sapiency.

The fullest fossil record of humans came from Asia and Europe and so Coon was most confident in his assessment of their evolutionary history. By contrast, Africa and Australia were poor in human fossils in the early 1960s. The fact was that the Negroid race seemed to arrive at the *sapiens* threshold last, although fossils discovered in the last few decades have suggested the opposite: that modern humans may have arisen first in Africa.[9] Australians, too, were seeming latecomers, and Coon noted that he had met individual Aborigines who were fully human and functioned normally in their society despite having a relatively small cranial capacity.

Coon also knew that this synthesis of the evidence and its bearing on both the past and the future might change when new material was discovered. He cautioned that the fossil record from Africa was scanty, compared with that from Asia and Europe, and predicted that the African picture might need to be revised or reversed "as new evidence becomes available."[10] Indeed, Coon was delighted years later when very old, large-brained skulls began to be found in Africa, a reaction which suggests he had no emotional commitment to the ill-founded myth that the Negroid race is intellectually inferior to any other race.[11]

Since the burning problem in America in the 1960s was one of prejudice against "colored people" or "Negroes"—terms soon deemed offensive and rapidly replaced by "blacks," then "Afro-Americans" and "African-Americans"—the crucial point was Coon's observation that fossils identified as Negroid were late in achieving modern brain size. This was taken by many as a slur against the Negroid race and evidence that Coon believed them to be intellectually inferior—which may or may not have been so. Coon attributed the "cultural dominance" and conspicuous achievements of Caucasoids and Mongoloids to environmental factors, to having been in the right place at the right time. It was a reworking of older ideas about the challenging nature of colder climates as opposed to the "easy" life in the tropics. He recognized that the world order was starting to change as evolution continued to proceed, writing:

[Caucasoids and Mongoloids] achieved all this because their ancestors occupied the most favorable of the earth's zoological regions, in which other kinds of animals also attained dominance during the Pleistocene. These regions had challenging climates and ample breeding grounds and were centrally located within continental land masses. . . . Any other subspecies that had evolved in these regions would probably have been just as successful. Now the success of these groups is being challenged in many parts of the world as other groups who evolved later learn to use their inventions.[12]

Was he indicating that he believed Caucasoids and Mongoloids are inherently superior intellectually to the other races? Perhaps. One of the oft-quoted passages in his book reads: "Wherever *Homo* arose, and Africa is at present the likeliest continent, he soon dispersed, in very primitive form, throughout the warm regions of the Old World. . . . If Africa was the cradle of mankind, it was only an indifferent kindergarten. Europe and Asia were our principle schools."[13] It is a passage that suggests no high regard for the intellectual accomplishments of Africans. Yet he persisted in defending the unpopular position that the relative intelligence of the human races was a matter about which there were insufficient data; he refused to express an opinion on the matter even when criticisms were fiery. Coon's personal beliefs about intelligence and race were the crux of the outcry, yet inferring them is difficult, for even those who knew him disagree.

However, if Coon's scientific premise about the origin of modern humans is correct—and the theory of regional continuity, somewhat revised, is still a powerful and popular one—there is one inevitability that must be faced. The chances are good that one human population evolved an average brain size equal to modern humans somewhat earlier than the other populations, even as one was probably last. There was, probably, one group in which the *average* brain size happened to get bigger faster, the same as some populations *on average* reach puberty earlier (or later). And because the fossil record in any particular region preserves only a handful of individuals from most periods of time, it is most unlikely that the fossils would appear to document exact simultaneity of this evolutionary development, even if simultaneity occurred.

Ironically, the issue of being the first to evolve average, modern brain size is a straw man. It is a development of no real import. Because there is a huge degree of overlap in brain size among all populations of one species at any given point in time, the significance of the "first" is nil. Besides, brain size has no demonstrated relationship to intelligence in any case. Coon cannot be blamed for reporting factually what was known of the fossil record, and it was bound to make one race or another look retarded. Where he can be faulted is for allowing his rather nineteenth-century self-assurance to push him into slithering out of the factual reality that, according to the evidence available to him, the Mongoloid race was first to reach sapiency.

As Coon predicted, new finds that improved the fossil record have provided a substantial revision of the sequence of events. As of this writing, the *earliest* large-brained fossils are known from the African continent, not the latest. This change in the evidence has produced a complete about-face in the details of the regional continuity theory and has thus brought it into the realm of the liberal and politically correct. Nonetheless, reputable and well-respected scholars still argue (as Coon did, and Weidenreich before him) that the constellations of physical traits that we recognize as racial grouping may have begun to develop in *Homo erectus* times.[14]

Whatever Coon's private views, his public statement on the *Origin of Races* caused an unparalleled application of blistering criticism to both his personal and professional personas. Fair or not, he was blamed for his account of the fossil record and for Putnam's use of it.

Much disapproval was directed toward Coon by Dobzhansky, who had months before sent his own book to Coon with a charming inscription. He lambasted Coon in his presidential address to the AAAS. He was also solicited to review *Origin of Races* by *The Saturday Review*, and he attempted to use this public forum to the fullest. However, his comments were so highly critical, even scathing, about Coon's grasp of evolutionary theory and his competency as an anthropologist, that *The Saturday Review* declined to publish Dobzhansky's words. Some extracts from Coon's book and a review by Margaret Mead appeared instead. Frustrated, Dobzhansky sent copies of his review to Mayr and Simpson, who were quoted favorably on the jacket of the book, as if to ask them to denounce what they had already endorsed. He eventually persuaded *Scientific American* to publish his review in February of 1963; it was also reprinted in *Current Anthropology*, the journal published by the Wenner-Gren Foundation. Montagu remembers that Dobzhansky showed him a letter from Coon, after the review appeared, threatening a lawsuit for defamation of character.[15]

Dobzhansky accused Coon of providing "grist for racist mills." While he would not advise a scientist to "eschew studies on the racial differentiation of mankind, or examining all possible hypotheses about it, for fear that his work will be misused," Dobzhansky also maintained that a scientist "cannot disclaim all responsibility for such misuses."[16] It is difficult to see what tack he felt Coon ought to have taken.

This was only the beginning of a river of disapprobation that Dobzhansky directed toward Coon on every possible occasion. At the last minute, Dobzhansky added an addendum to the published version of his paper for the Wenner-Gren conference "Classification and Human Evolution," a meeting to which Washburn had not invited Coon. Dobzhansky obviously knew the volume would be widely read and felt it was an appropriate arena in which to express his opinion of Coon's book. Of Coon's argument that *Homo erectus* was transformed not once, but at least five times, into *Homo sapiens*, Dobzhansky remarked:

> This belief, which has made Dr. Coon's work attractive to racist pamphleteers, is neither supported by conclusive evidence nor plausible on theoretical grounds. The specific unity of mankind was maintained throughout its history by gene flow due to migration. . . . Excepting through such gene flow, repeated origins of the same species are so improbable that this conjecture is not worthy of serious consideration; and given a gene flow, it becomes fallacious to say that a species has originated repeatedly, and even more fallacious to contend that it has originated five times, or any other number above one.[16]

Coon was livid. He cornered the young Lita Osmundsen, new in her role as director of the Wenner-Gren Foundation, the next time they met. Looming over her, he inquired fiercely how she could have permitted such a scurrilous addition to the conference volume. Undaunted, she pointed out that the foundation had sponsored the research behind his book, without ever telling him what to say, and they weren't going to tell Dobzhansky what to say, either. He glowered at her for a moment, and then his face broke into a grin. "Girl," he said gruffly, "you're all right."[18]

Dobzhansky's acrid jibes at Coon were not confined to the professional sphere. He pointedly refused to shake Coon's hand or even to acknowledge his presence at scientific meetings; he walked past Coon, looked through him, turned his back on him, and generally transformed a scientific disagreement into personal rudeness. His censure continued up until early 1968, when Dobzhansky published a piece called "More Bogus Science" in the *Journal of Heredity*. The article purported to be a review of Carleton Putnam's second book, *Race and Reality*—which was as racist as the first—but much of the text was yet another attack on Coon's book. Dobzhansky insulted Coon's theory as being "erroneous," "far-fetched," "typological," and, on particular points, "arbitrary" and "invalid"; it had also been "rejected by most of his colleagues, eminent and simply competent ones" according to Dobzhansky. Coon's worst sin, in Dobzhansky's eyes, was in not preventing or rejecting Putnam's use of his findings: "It is the duty of a scientist to prevent misuse and prostitution of his findings," Dobzhansky moralized.[19]

Coon replied furiously in a letter to the editor of the journal.

Dobzhansky was misrepresenting Coon's theories, either through care-lessness or intent, and their reception by his colleagues, Coon assert-ed. More to the point was the issue of a scientist's moral responsibility. Coon fumed:

Had Mr. Putnam misquoted me I would have said so long ago. Dobzhansky states that 'It is the duty of a scientist to prevent mis-use and prostitution of his findings.' I disagree with him. It is the duty of a scientist to do his work conscientiously and to the best of his ability, which is exactly what I have done and shall contin-ue to do, and to reject publicly only the writings of those persons who, influenced by one cause or another, have misquoted him, as Dobzhansky repeatedly has done with my work, for reasons best known to himself.

Were the evolution of fruit flies a prime social and political is-sue, Dobzhansky might easily find himself in the same situation in which he and his followers have tried to place me.[20]

The bitterness was pervasive. Dobzhansky was not the only one to criticize Coon for what he *did not* say. A letter written in 1964 by Sid-ney Mintz, a social anthropologist at Johns Hopkins University, was quoted recently in *The New York Times* by Ashley Montagu. Mintz nev-er intended the words for publication[21] and says now that he would have moderated the tone had he known his words would appear in print; still, he stands by the sentiment: "Coon himself has carefully refrained from following up any of the implications of his argument. My own convic-tion is that he is a racist of the worst sort, but extemely clever. He has left the dirty work to others."[22] What Mintz really faulted Coon for, then and now, was failing to repudiate Putnam's brief reference to Coon's work for his own ends. Reading the caustic exchanges between Coon and his attackers, there is more than a hint that the critics were so en-raged that no denial or avowal on Coon's part would have turned aside their wrath.

It was an unresolvable conflict between the fervent social activist and the irascible scientific purist. But the tenor of the times was such that it was the scientific purist, Coon, who was disgraced and, to some extent, driven out of his profession. Months before Dobzhansky's death in 1975, Coon tried to make peace in a letter; Dobzhansky did not an-swer. In short, for thirteen years Dobzhansky remained dogmatic in his complete condemnation of Coon's work and person.

Dobzhansky was not alone in his vilification of Coon. On November

16, 1962, Dobzhansky's friend Sherwood Washburn stood up in front of the annual meeting of the American Anthropological Association to give his presidential address. He began by asserting that his choice of topic was urged on him by the executive board of the association and he had reluctantly agreed, as if to disclaim responsibility for what would follow. And then, in the words of one of Washburn's former students, "Sherry tore into Carl like you wouldn't believe."[23] It is an event that has been incorporated into the mythology of physical anthropology as the great public denunciation and humiliation of a once well respected man. Washburn, too, poured forth a stream of accusation and abuse.

The published version of the address, and the vivid memories of those present at the time, are the only record of exactly what was said. According to Coon, Washburn initially tried to prevent publication of the speech, perhaps realizing that he had gone too far.[24] In any case, Washburn mentioned Coon or his book by name only once in the published version—and then only in most unfavorable contrast to Dobzhansky's *Mankind Evolving*, which had been published only a few months previously. Washburn's veiled references to Coon's work did not conceal the identity of his target from the audience. It was a scandal in the field.

Washburn spoke obliquely of Coon's work as typological and old-fashioned, feigning amazement that "this kind of anthropology is still alive . . . and in full force."[25] He also ridiculed Coon's theories about the possible adaptive value of facial and nasal shape in different races and fossil species, although viewing morphology in terms of evolution and adaptation was exactly the sort of perspective Washburn's New Physical Anthropology promoted. Indeed, Coon's effort was an important milestone in this regard. Washburn labeled Coon's ideas as "nineteenth century . . . an extraordinary retrogression to the worst kind of evolutionary speculation—speculation that antedates genetics and reveals a lack of any kind of reasonable understanding of the structure of the human face," and insulted Coon as being "anatomically illiterate."

In his speech, Washburn also denied the relevance of race and race studies in modern science.

> Since races are open systems [genetically] which are intergrading [into each other], the number of races will depend on the purpose of the classification. This is, I think, a tremendously important point. . . . Race isn't very important biologically. . . .

Race, then, is a useful concept only if one is concerned with the kind of anatomical, genetical, and structural differences which were in time past important in the origin of races. Race in human thinking is a very minor concept. . . .

When the meaning of skin color and structure is fully understood, it will help us to understand the origin of races, but this is not the same thing as understanding the origin of our species. It will help in the understanding of why color was important in time long past, but it will have no meaning to modern technological society.[26]

Washburn could not be ignorant of the urgent and painful drama—the sit-ins, the protests, the violence, and the unquenchable hope—being played out throughout America that turned on just such questions as skin color and race. The assassinations of the Kennedys and Martin Luther King, Jr., were yet to occur, but the vehemence of the struggle for racial equality was patently obvious even in the early autumn of 1962. Washburn probably would have preferred race to be irrelevant to society, as many would, but this was far from being true. His solution was to downplay and denigrate scientific study of racial differences, thus hoping to rob the racists of scientific credibility for their arguments. In retrospect, it might have been wiser to urge even more study, because any deep understanding of the intricate pattern of variation of traits among humans can only underscore our unity.

In closing, Washburn turned to the ever-incendiary issue of racial differences in intelligence, asserting that there was no evidence that the reported differences in IQ between blacks and whites reflected differences in genetic ability rather than in environment. He and Coon agreed on this point, though Putnam did not, but the audience had already grasped that Coon was Washburn's primary target and may not have understood the distinction here.

What was clear was that any anthropologist who dared to deal with race was risking a similar mortification at the hands of the powerful and influential. Washburn's address wounded Coon deeply. Instead of retiring with honor, as a grand old man of anthropology, Coon went off, head high and back stiff, but disgraced in most of his colleagues' eyes.

Coon was stubborn and was no quitter. Some months later, he decided to use his second presidential address to the American Association of Physical Anthropologists, who were meeting jointly with an archaeological association, to answer Washburn's attacks "all in good humor."[27] No one knows what Coon would have said. On the opening day, a colleague of Washburn's got wind of Coon's intention and spoke to the conference organizer, who decided to abort both presidential addresses and speak himself, on a less fraught subject.

Others jumped into the fray.[28] There was a letter of protest against Coon's book in *The New York Times* written by Henry Schwarzchild of the Anti-Defamation League. Morton Fried, an anthropologist at Columbia University, circulated a flyer to eighty-four teachers of anthropology who he thought might be considering using Coon's book as a text. He and Dobzhansky were hoping to set up a committee to raise money to place advertisements in various newspapers, apparently to discredit or disown Coon and Putnam together. The effort was apparently abandoned without success.

Fueled by his own internal fire, Ashley Montagu wrote a strongly worded article in *Current Anthropology*, again attacking Coon's book. Montagu singled out various words and phrases of Coon's that seemed to indicate an old-fashioned or racist sentiment. He was sarcastic about Coon's description of the five subspecies of *Homo erectus* as "more brutal" than their modern descendants. " 'Brutal' is a word capable of several meanings in such a context, and when juxtaposed to 'sapient' perpetuates pejorative and odious comparisons which are of questionable validity in scientific or other discussions," Montagu observed.[29] He wrote with scorn of Coon's "fecund imagination" in conceiving of the parallel evolution of the five subspecies of *Homo erectus* into *Homo sapiens*.[30] He cast aspersions on Coon's illustrations, photographs of people of different races in their usual attire. Montagu focused in on Plate 32, which shows an Australian Aboriginal woman at the top of the page and a Chinese man at the bottom. Coon's caption read:

> The Alpha and Omega of *Homo sapiens*: An Australian aboriginal woman with a cranial capacity of under 1,000 cc (Topsy, a Tiwi); and a Chinese sage with a brain nearly twice that size (Dr. Li Chi, the renowned archaeologist and director of the Academy Sinica).[31]

Montagu was incensed by these words.

> How does Dr. Coon know that Dr. Li is a sage [Montagu protested]? Does being an archaeologist make him so? Or does the apposition of "aboriginal" and "sage" serve to make a point? "Alpha and Omega," the first and the last. "Obviously" Topsy just growed [this was a reference to a character named Topsy in *Uncle Tom's*

Cabin] and is what she is, a poor benighted Australian Aboriginal, principally because she has a small brain, and Dr. Li is principally what he is because he has a large brain. Of course, there are cultural differences, but the implication is that no matter what cultural advantages Topsy and her children had been afforded, neither she nor they could have accomplished what Dr. Li has achieved.

. . . It is surely both unfair and unscientific to compare cranial capacities of two individuals from two different groups, and then proceed to draw conclusions from such a comparison for the groups as a whole. The citation of averages and their statistical variance would have been somewhat more meaningful, even though that meaningfulness would be limited to the clearer statement of the variation in cranial capacity characterizing the two groups.[32]

In his fury, Montagu put words into Coon's mouth that were more offensive than what he actually said, which was at best insensitive. Montagu was probably also reacting to the dreadful irony of the woman's name, which smacks of a cruel white man's (or woman's) joke—but Coon cannot have named her himself and ought not be blamed for this travesty. On Coon's part, to call a fellow archaeologist a sage was simple courtesy, and apparently the external measurements of his head indicated a large capacity. It is also likely that Topsy, pictured in her traditional Aboriginal dress, was not likely to have been an educated woman and may, indeed, have had a somewhat small cranial capacity—as many women and many Australian Aborigines do.

But the crucial issue is that neither brain nor skull size has any relationship to the intelligence of the individual. Perhaps Coon only meant to demonstrate the range of variability within modern humans, but labeling one Alpha and the other Omega seems part of the pervasive nineteenth-century habit of dehumanizing and denigrating "natives." The pity is that Coon outlived the era in which his contributions might have been most valuable, in which he would have been happiest; the tragedy is that his honest attempt to wrestle with a major problem was misused by racists and abused by liberals. It is impossible to discern what Coon intended to imply about larger racial issues with this caption; it is only possible to be certain that Montagu perceived a very ugly implication indeed. Coon's reply to Montagu was so heated that the editor of *Current Anthropology* hastily declared the correspondence closed.

But it was not. The burning nastiness burst forth again in 1965, when Coon published the second and final volume of his life's work, entitled *The Living Races of Man*. Trying to avoid controversy, Coon prefaced his book with a quotation from Poincaré, to the effect that here were the simple facts, which were not subservient to dogmas, passions, or partisan causes, for such submission would destroy them. Coon also altered the scope of this book from his original plan.

> In the introduction to [*The Origin of Races*, Coon wrote] I stated my intention to discuss in this one racial differences in blood groups and the anatomy of the brain, but only blood groups have been covered. Racial differences in the brain imply differences in intelligence, a subject so laden with emotion that its mere mention evokes unsolicited acclaim and feverish denunciation. Even without reference to the brain or to intelligence, the simple statement that races exist drives a small coterie of vocal critics into a predictable and well-publicized frenzy.
>
> I hope but do not expect that all reviewers will read the whole book and not just the introduction and the final chapter. I also formally request that no one shall quote or cite this book as ammunition for or against any cause or movement whatsoever, because, as any reader can see, I have tried throughout to adhere to the principles clearly stated in the preceding quotation from Henri Poincaré. If anyone quotes, or denounces the book or myself because of some fancied adherence to any dogma, cause, emotion, personal interest, or preconceived idea, I shall conclude that he cannot read simple and beautiful French, or even plain English.[33]

It was to no avail. His words and discretion boomeranged. Ads for the book called it "the most controversial report on *Homo sapiens* since Darwin," and stated that Coon "reaches some startling conclusions that pull all the myth-balanced props out from under the racists."[34] Some reviewers, predictably, took the opposite view. Coon was again criticized for his photographs of representatives of different races, by the prominent British anthropologist Edmund Leach, because "the *Caucasians* are posed in shirt-sleeves and 'civilized' haircuts, whereas most of his other categories appear as bare-arsed savages."[35] As Coon pointed out, the alternatives were either to dress the tribal peoples in unaccustomed garments, after cutting their hair, or to photograph the Caucasians "bare-arsed."

Alice Brues, an anthropologist who also reviewed the work, praised Coon.

> It requires a degree of courage to write a book on the Races of Man in this era of the New Prudery, when r-ce has replaced s-x as the great dirty word. Some of the criticisms which the work will receive reflect an adrenergic reflex to the title itself. And paradoxically, criticism will be more severe because there is nothing else like it in print and people will have to refer to it whether they agree with it or not. . . .
>
> A refreshing feature is the photographic section, which shows 130-odd individuals, enough to include not only "type" specimens of major races but a good range of variation within races. In contrast to anthropological illustrations of the older tradition, in which the subjects were too often shown in poses painfully reminiscent of the official record of a convicted felon, Coon's pictures show pleasant, interesting human beings, their dignity preserved by showing them in natural pose and expression, and clothes, if such is their usual custom.[36]

Brues was exactly right: *race* was by 1965 a dirty word, a taboo subject for anthropologists. Only a pigheaded man like Coon would have dared to write about it. Like an explorer determined to conquer a volcanic peak, he could not or would not remove himself from danger and so he was engulfed.

Leigh Van Valen was one of the few evolutionary biologists to spring to Coon's defense, and to criticize Montagu and Dobzhansky for their harsh words. Van Valen was, at the time, a research fellow at the American Museum of Natural History. His analysis of the controversy is both pungent and perceptive, but it was rejected for publication by *Current Anthropology* on account of the journal's "peculiarly intense discussions of these issues in the past."[37] Van Valen published instead in a more obscure, less widely read journal, *Perspectives in Biology and Medicine*, where—sadly—it had relatively little impact on the continued course of the controversy.

Van Valen's most telling points were that Coon's theses were irrelevant to modern racism and that the truth or falsehood of Coon's arguments cannot be judged on their political consequences but must be evaluated paleontologically.

Whether one race or another reached sapiency last, Van Valen observes, historical precedence in braininess can hardly be used to justify discrimination against one group or another now, because of the enormous overlap in all measures of intelligence among living human groups. Nor, he argues, can the (potential or actual) misuse of facts or theories by racists excuse a scientist for rejecting a line of evidence suggestive of racial inequality *solely* on the grounds that the conclusion is politically unwelcome.

> Misuse of the principle of "colorblindness" by racists should not force us into the politically less ambiguous, but now still scientifically untenable, position of denying the possibility of average racial differences in mental characters. To ignore race and treat an individual as an individual is the spring of justice and the river of hope. This attitude does not exclude the use of special remedial measures to reduce the effect of economic, racial, or other discrimination; to say otherwise is equivalent to denying medical treatment to the sick.[38]

He includes a long discussion of different genetic mechanisms whereby Coon's hypothesized evolution of geographical populations of *Homo erectus* might evolve in parallel toward *Homo sapiens*—as groups sufficiently distinct to have recognizable racial characteristics without being so isolated that no gene flow occurred. To Van Valen, this idea is certainly plausible and falls within the well-established tenets of population genetics.

At the end of the paper, he outlines the crucial distinction between matters of scientific fact and their political implications:

> These severe distortions, amounting to caricatures, given by Dobzhansky and Montagu of Coon's clearly presented argument are presumably not deliberate, but they are indicative of a reluctance to face objectively the possibility that different subspecies of *Homo sapiens* did arrive at this self-exalted grade at different times. . . . The proposed mechanism is plausible, and even if not it would be irrelevant to the substantial and factual questions of whether the extant geographical races of man are detectable in middle Pleistocene fossils, and if so whether they arrived at the grade of *Homo sapiens* at different times. These quesions are paleontological and can be answered only by further study and increase of the paleontological evidence.
>
> A second and equally important piece of evidence on the prejudice of these critics is their emphasis on the proper (Montagu)

or improper (Dobzhansky) use of Coon's theses for racist ends, by Coon himself (Montagu) or others (Dobzhansky). To condemn a scientific enquiry because of its possible political consequences is bigotry, no matter by whom practiced. . . . Coon's theses are irrelevant to racism. . . . But whether they in fact support racism is a question that is utterly distinct from whether they are true. To confound these points is to evaluate a question of truth, presumably divinely revealed because not open to question, by a criterion of value or utility. And this is to padlock the mind.[39]

Van Valen's prose is strong and bravely delivered. But even he felt the need to protect himself by establishing his liberal credentials. In the only footnote to the paper, Van Valen declares: "It should not be necessary, but probably is, to say that I have been associated with civil rights groups for half my life and have initiated several desegregation campaigns."[40] It is a sad commentary on the state of politics and science.

Too few read Van Valen's words in 1966 and Montagu, if he read them, did not heed them. Fatefully, Carleton Putnam chose to publish another racist book in 1967, aggravating the situation still further. In 1969, Montagu denounced Coon on a television talk show and Coon believed his painful trials were over. He died in 1981, unredeemed in the eyes of the anthropological community; his precious theory about the origin of races seemed to be in eternal disrepute. His obituary in *The New York Times*, on June 6, 1981, read in part:

Dr. Carleton S. Coon, one of the last of the great general anthropologists, died Wednesday at his home in Gloucester, Mass. He was 76 years old. . . .

A deep interest in the origins of race led Dr. Coon to expound a theory that five major races of man differentiated even before the emergence of *Homo Sapiens* as the dominant human species. This theory was never widely accepted by scientists and is now largely ignored. Dr. Coon's theory was sometimes used by racists to support their views, but he explicitly repudiated their contentions in the second edition of "The Story of Man" in 1962.[41]

This last point—that Coon repudiated the use of his work by racists—was information presumably supplied by Coon's widow or family. He may have felt he had dissociated himself from racist claims, but his critics did not agree. Coon had wrestled with the painful subject of evolution of human races and had come away covered in burning welts that turned to lasting scars. He never regained his prominence in his field.

But the criticism has continued posthumously. In 1992, Montagu derided Coon's work in a letter to *The New York Times*. The occasion was a discussion of an anti-Caucasian work, *The Iceman Inheritance*, by an African-American author, Michael Bradley.[42] One of Bradley's contentions was that Caucasians were "icemen" descended directly from Neandertals, a heritage that makes Caucasians more aggressive and vicious than Negroid peoples, whose origin was different according to Bradley. Human paleontologists find this reading of the fossil record incredible but, ironically, Bradley cited Coon's *Origin of Races* as the authority for this statement. This shows how rich Coon's book was in data that could be interpreted flexibly.

Montagu is as unyielding as ever in his remarks:

> Michael Bradley [says that his] book was merely quoting the conclusions of "The Origin of Races," a 1962 work by Dr. Carleton Coon.
>
> The findings of the Carleton Coon book were neither scientific nor sound, and its conclusions were rejected by physical anthropologists as both genetically ignorant and socially prejudicial.
>
> Among Coon's many erroneous conclusions, for example, was the claim that the "Negroid race" evolved to a sapiens species status 200,000 years after the "white race." The inference was clear. Since, according to Coon, "Negroes" were the last of the Homo erectus subspecies to be transformed into Homo sapiens, their level of development and civilization is "explained." Negroes, he argued, simply do not have as long a genetic history as "whites," and this accounts for their "backwardness."[43]

Montagu has here put in quotes words that are not actually Coon's but are paraphrases of what he reads Coon to mean. For example, Coon consistently refers to the races as Caucasoids, Mongoloids, Australoids, Capoids, and Congoids (a group including African pygmies and Negroes, in Coon's usage) rather than by tags based on skin color, and he does not refer to the "backwardness" of Congoid or Negroid peoples of the world. In conclusion, Montagu says, "It is unfortunate that Mr. Bradley chose Coon's book as the authority for his views."

Was any good engendered by this bitter and dishonorable episode in the history of the long argument over human evolution? Certainly sensitivities were raised in the anthropological community to any remarks or references that might be seen as denigrating an ethnic or racial group. But what was produced was more than an allergic reaction; the response approached anaphylactic shock in seriousness and nearly killed off the study of race for good.

Those who want to resolve modern society's turmoil by terrible means should be found publicly guilty when they subvert science to their convictions. Such people are ever too ready to hear any faint whispers of support for their racist beliefs, too eager to pick and choose their facts to fit their theories, without any commitment to rigorous evaluation of those "facts." The scientific community can neither condone such acts nor can they be lax in pointing out the logical and factual flaws in the work of racists. In this case, the tragedy is that the perfidy did not stop there. Those who meant to strike blows for equality and freedom also bent the truth to their cause. They, too, misquoted or misrepresented others' work, failed to evaluate evidence objectively, or blamed others for what was not said. Their fervor inhibited any rational or scientific discourse on the subject of race and simply compounded the error.

And the enigma in the middle, Carleton Coon, was scarred by both sides. He aimed to present the facts, to provide a synthesis, to outline a clear hypothesis that could be tested against new fossil and biological evidence as it became available. He did a fair job of accomplishing his aim, though he—like nearly every scientist in a field where data are scarce—may have overinterpreted his evidence. But pushing the data, stretching for the synthesis, is exactly what theory-building requires. Only a *testable* theory or hypothesis is of any use; only once a theory has been articulated clearly can it be tested against new and more sufficient evidence and perhaps refuted. Coon constructed an explicit hypothesis and did not fail to draw clear lines around matters of fact and matters of speculation in his book. Whatever lurked in his heart, be it the darkness of prejudice or the light of the love of fellow humans, he performed a valuable and courageous service.

Race was simply too volatile a subject, political exigencies too compelling, for a disinterested treatment. Race became the dirtiest word in the lexicon and the entire next generation of anthropologists were frightened away from dealing objectively with a real and urgent issue. The argument could no longer even be held in public by the scientists who, by training, were the best qualified to discover the facts and debate their implications sanely. Race and human variability simply disappeared from the curricula in most colleges and universities in the English-speaking world. Rather than educate the next generation about what race is, and isn't—rather than trying to develop an informed appreciation for the diversity and richness of the traits that people inherit from an admixture of the now-blurred races—the upshot of the attacks on Coon was to deny that the subject even existed.

But the problems of multicultural societies did not disappear with Coon's defeat. There was no greater harmony among people of different appearance and ethnic background, especially where conditions of poverty and indifferent education exacerbated the tensions. There was no diminution of prejudice, name-calling, and slashing attacks that relied on false facts about the genetic capabilities of various groups. If anything, the widespread refusal to acknowledge, much less examine, racial issues intensified the problems. From the anthropological community, there was only a profound and dangerous silence.

In April of 1993, at the annual business meeting of American Association of Physical Anthropologists in Toronto, a motion was made to approve an updated version of the UNESCO document on biological aspects of race. Its preamble shows the legacy of Washburn, Montagu, and Dobzhansky:

> As scientists who study human evolution and variation, we believe that we have an obligation to share with other scientists and the general public our current understanding of the structure of human variation from a biological perspective. Nineteenth and early twentieth century categories of race have often been used to support racist doctrines. The race concept currently is understood to have little scientific merit when compared with other biological and social units of analysis, and persists as a social convention that fosters institutional discrimination. The expression of prejudice may or may not undermine material well-being, but it does involve the mistreatment of people and thus it often is psychologically distressing and socially damaging. Scientists should endeavor to prevent the results of their research from being used in a biased way that would serve discriminatory ends.[44]

The body of the draft statement also reveals its evolutionary ancestry. Its eleven brief points emphasize the unity and value of all living humans and assert the nonexistence, now or in the past, of pure races; they propose that genetic, natural, and social environments influence the physical differences among humans; they observe that human patterns of constant migration and interbreeding have produced a complex pattern of diverse traits that, even when heritable, are rarely linked to each other and are not tied to cultural, linguistic, or geographical groups.

All of these points are undisputed and true, or as close to true as modern science can determine. But the exact wording of documents like this is incredibly important; a wise association, before endorsing such a statement, scrutinizes the nuances and implications of each passage. The motion did not pass; the document was sent back to the committee for revision. This outcome is not surprising. It has been said, not entirely in jest, that the number of opinions in a room full of anthropologists is roughly equivalent to the number of anthropologists. What is surprising is that few engaged in the discussion seemed aware of the history of UNESCO statements on race and the furor they raised.

Let us hope that those who are ignorant of history are not condemned to repeat it. Let us hope that something has been learned about civility and tolerance.

Part VI
The Genetics
of Racism

The Ratio of
Light to Heat

CHAPTER 13

HE NEVER DEBATED documents on biological aspects of race. He didn't even know he was writing about race. But his ignorance—or naivete—did not shield him from the inevitable controversy. His name is David Wasserman.

Over the past thirty years, anthropologists' unwillingness to address the vexatious problem of human racial differences has shifted the arena for the debate from the academic to the public realm. The application of evolutionary theory to humans became such a commonplace that it is the nearly invisible background of much of modern biological science. Genetic detail has deepened and enriched neo-Darwinian evolution, but it has resolved nothing about human races and the value of the differences among them. The arguments have simply became more heated and less tractable, for they are no longer the property of objective scrutiny. Following genetics came molecular biology, which has tended to reduce organisms to collections of inanimate biochemical reactions. Reductionism has bred a terrible sense of individual powerlessness—that no one was responsible for his or her attributes or even exercised control over them. It was all just molecules, mindlessly binding or releasing, attracting or repulsing. Not only evolution got lost; humans got lost, too.

If there is a man whose recent history has been marked by this new twist in evolutionary science, it is David Wasserman.

Long after the brouhaha over Coon's *Origin of Races*, Wasserman witnessed a controversy that was in many ways a repeat of Coon's experience. Wasserman thought he was organizing a conference that would focus on the cutting edge of modern genetics, on the interaction of a new technology for exploring human differences and the American legal system. And he didn't see that the seemingly innocuous inclusion of a topic that mixed behavior with genetics would poison his conference and tarnish his reputation.

Unlike Carleton Coon, Wasserman was no scholar of race and racial differences; he was unsuspecting of its powerful dangers. Indeed, Wasserman came from a personal and professional background that might have rendered him invulnerable to accusations of racism. Raised in suburban Connecticut, Wasserman is the child of two liberal, well-educated parents. His physician father is described by Wasserman as "a paragon of the virtuous private practitioner"; his mother, with a master's degree in education, is on the Board of Governors of Higher Education in Connecticut and is "one of the last of the great woman volunteers."[1] Wasserman was raised with a heritage of public concern and activism, as well as a minority status as a Jew. A bright young man, he attended Yale University as an undergraduate, majoring in philosophy, then went on to the University of Michigan Law School, where he earned a J.D., and to the University of North Carolina for a master's degree in social psychology.

This eclectic training suited his interests well, for he has always been engrossed by the issues at the complex interface of philosophy, law, and public policy. How does society work—and fail? What should be done to heal, to unify, to improve? His is the perspective of the knowledgeable optimist—a criminal appellate defender, rather than a prosecutor. After practicing law for some years, he yielded to his longtime fascinations and embarked on a career as a legal scholar and ethicist. His publications touch on an interesting mixture of topics: the psychology of decision-making by juries, self-defense, hindsight and causality, and moral issues in statistical inference. He also drew on his experience as an appellate defender to write a book on criminal appeals.

In the 1980s, as evolutionary theory migrated from the biologist's laboratory into the courtroom—via innovative applications of new techniques—Wasserman began to track the effects of these new technologies on the legal system: what happened when biological law met human law? In 1990, he won a National Science Foundation professional development grant to study the emerging field of genetic typing and criminal justice. His has not been simply an abstract study; in an effort to come to grips with the practicalities as well as the burgeoning moral issues, Wasserman tackled both course and laboratory work at the Human Genetics Division of the University of Maryland.

The premise behind the invasion of law by genetics is based on one of the earliest observations to come out of the new evolutionary synthesis: that all individuals vary in innumerable ways and are *genetically* unique (except for identical twins). In practice, this means that each person's genotype codes for a singular combination of polymorphisms, or the slightly different versions of the same proteins and enzymes produced by the body during development from the embryo and during its day-to-day functioning. Determining an individual's genotype—or part of it—is known as genetic fingerprinting, because (like a fingerprint) the pattern of polymorphisms in an individual is used as a means of identification. It was the advent of the polymerase chain reaction (PCR) machine, which is capable of cloning or reproducing many times over an individual's DNA from a minute sample of hair, blood, semen, or spit, that made it feasible to work with scene-of-the-crime evidence. Once DNA samples are obtained from two people—say, a suspect and a perpetrator of a rape—it is simply a matter of comparing two specific patterns of polymorphisms in their genotypes. If they are the same, so are the individuals. It is rarely feasible to examine the entire DNA sequences; fingerprinting relies on partial genotypes and on data that show how rare or common each particular polymorphism is in the population at large. These comparative data are essential background information. The more polymorphisms are examined, the better the chance of convicting—or clearing—a suspect.[2]

Darwinian theory has come a long way, from a natural history perspective on the evolutionary history of mankind to a high-tech, objective resolution of the identity of a criminal. In its first hundred years, evolution left the quiet meeting rooms of gentlemen's clubs and scientific societies and moved into the offices of politicians; in the past few decades, it has wandered from the sterile laboratories of high-tech geneticists into the courts and jails of modern society. Evolutionary theory is no longer about birds and bees, fossil and living species, alone; it is increasingly about humans and their social and medical diseases and disorders. If the examination of races and racial attributes became disreputable in the 1960s, the introspective tendency to apply neo-Darwinian theory to humans did not. And with the forging of an unbreakable link between evolution and human genetics, the problem of the value of the differences among humans rose to the fore in a new way.

Wasserman's focus was narrower. He didn't see the broader evolutionary issues or implications for racial differences; he simply wanted to know how these new techniques *worked* and what the dangers and drawbacks were for the legal system. Were the statistics—those ever-so-convincing figures like "there's only a one-in-a-million chance that someone other than the suspect could show this precise pattern"—accurate? He also hoped to explore the implications of a statistical identification, a *probability* of guilt: Were juries and judges sufficiently educated in probability theory to deal with such concepts? These were compelling questions for Wasserman.

In the same way, he was drawn to the new Human Genome Project, a mega-science endeavor that was commandeering the largest federal appropriation for a single project ever. The objective of the project is to sequence the entire human genome, or genetic sequence; this is an entirely feasible, if time-consuming, task requiring considerable technical expertise—it promises to fund labs full of postdocs and graduate students for years to come—but the sequencing itself demands relatively little intellectual prowess. The more challenging part comes *after* the sequence for some particular segment of the genome is obtained, as researchers struggle to identify genetic markers for various diseases, predispositions, and ailments, and to understand the processes involved. Already, everything from obscure diseases to predispositions to behavioral problems like alcoholism have been linked (tentatively or firmly) to specific genetic markers. Finding such a marker is widely regarded as the first step to identifying exactly what goes wrong, which biochemical pathway gets diverted from its normal course, or which protein or enzyme is either deficient or overabundant.

It is all part of a grand, new reductionist view that both disease and behavior are ultimately genetically controlled, a view that provokes troubling implications. In 1992 in the *Vanderbilt Law Review*, Dorothy Nelkin, a sociologist, and Rochelle Cooper Dreyfuss, a legal scholar, identified a disturbing tendency toward "genetic essentialism" in the courts.[3] This is a sort of modern-day predetermination that implies that an individual's behavior and medical history are set by the genes he or she inherits. The old Calvinist notion that the good or evil people do is predestined, their role assigned at birth by a higher and uncontrollable power, is sneaking back into the public consciousness, with the telling substitution of the Almighty Gene for God. Social, economic, and environmental factors that might influence either health or behavior are being downplayed; according to Nelkin and Dreyfuss, this is an attitude typical of periods in which the economy is bad and people reluctant to spend money on massive social programs.

What's more, genetic essentialism is one way of absolving individuals of personal, ethical responsibility for their actions. Who can be blamed for stealing, any more than one can be blamed for getting sick, if both are in his or her genes? Why should a parent accept responsibility for the difficult task of instilling discipline and moral values in a child if these are inherited? The logical extension of these arguments is to question whether crime is "bad"—morally reprehensible—at all; perhaps it is just "doing what comes nat'rally," as the song goes. Genetic essentialism is the beginning of a slippery slope indeed. And because different human populations or races differ in the frequency of various genes, racism and prejudice find this downhill run inviting.

For Wasserman, the decisive issues are how society reacts to this increasing reductionism of behavior to biochemistry and how this new type of information will change the legal system. As more and more claims are made for the underlying genetic component at work in antisocial behaviors, where will the lines be drawn? How will our notions of responsibility and our practices of blaming be affected if criminality is seen as genetically controlled? Is crime a *medical* not a social problem? Wasserman's background in statistics and his newfound understanding of the laboratory techniques involved also make him wonder about standards of evidence and proof: What would constitute a meaningful correlation between a gene and a behavior? The air of impenetrable technical complexity, impartiality, and utter certainty with which DNA fingerprinting and other scientific methods are presented to laypeople has been dispelled for Wasserman by his fellowship studies; he knows enough to know there may be spurious associations, false-positives, and other causes of doubt. So what demands for rigor and replicability ought society impose on genetic marker studies?

Wasserman saw his interests as a perfect fit with the new program to fund research exploring the ethical, legal, and social implications of the Human Genome Project. He envisioned, and began to assemble, a proposal for an exciting conference that would tackle these important questions. Wasserman predicted that, as the Human Genome Project proceeded, sooner or later someone would claim to have identified a genetic marker for some component of criminality or for a biochemical tendency—for example, a greater sensitivity to particular neurotransmitters—that might make an individual more "impulsive." The new technologies that made the Human Genome Project possible enabled such a finding.

"But," as Wasserman observed in his grant proposal,

genetic research also gains impetus from the apparent failure of environmental approaches to crime—deterrence, diversion, and rehabilitation—to affect the dramatic increase in crime, especially violent crime, that this country has suffered over the past 30 years. . . .

The [future] discovery of genetic markers associated with crime may encourage scientists and policy-makers to overlook the enormous complexities in any genetic contribution to criminal behavior. Medical geneticists have found that very similar disease phenotypes are often produced by different genotypes . . . ; while the same genotypes often produce different phenotypes. . . . The complications are even greater in studying criminal conduct, which is not only socially elicited but socially constructed: defined by legislatures and courts and detected by arrests, convictions and self-reports. . . . Genetic contributions to such culturally defined and ascertained conduct will be difficult to track and easy to oversimplify. . . .

It is probable that our ability to forecast will outstrip our ability to explain: we may detect criminal predispositions through family history, linkage and segregation studies, enzyme assays and gene probes well before we understand the genetic contribution to crime. . . . Because of the gap between prediction and understanding, a little genetic knowledge may indeed be a dangerous thing.[4]

The point of the conference, he explained, was

to address the impact of behavioral genetics on the criminal justice system. . . . We have found strong interest in, and a strong perceived need for, a conference like ours, on the part of scientists and legal scholars alike: the former are concerned with integrating the findings from different fields and research strategies, and in preventing the distortion and abuse of those findings by the legal system; the latter are anxious to learn about the type of findings the legal system will confront and to consider how it should receive them. . . .

In the past the criminal justice system has been a voracious consumer of both good and bad scientific research. It has often, to adapt G. B. Shaw, gone from uncritical acceptance to wholesale rejection without passing through an intervening stage of cautious use. . . .

In sum, we hope to improve the ratio of light to heat in a debate that will only intensify as findings about the heritability and neurobiology of violent and antisocial behavior proliferate, and as proponents of genetic explanation become more ambitious.[5]

Already there was a long tradition of evolutionary research into the heritability of criminal behavior—work that traced its ancestry through Sheldon's constitutional research, various studies of the infamous Jukes and Kallikak families, and back to Cesare Lombroso's phrenology of the criminal type in the nineteenth century. So Wasserman invited some of the best modern proponents of heritability studies to make their arguments that criminality seemed to run in families regardless of social status, economics, or education. But he also wanted the geneticists to come, both those who thought there might be a genetic component to criminal behavior and those who were the foremost skeptics of this idea. And he deemed it crucial to include criminologists, legal scholars, sociologists, philosophers, and ethicists in the mix, as well as historians of science who were experts in eugenics.

The impressive array of scholars that Wasserman contacted were to accomplish a novel task: instead of talking about research that was already completed and published—the usual scientific procedure—Wasserman wanted them to wrestle with concepts, validity, and implications *in advance* of the scientific work. This difference in perspective, in the philosophical tradition of discussing what *might* come rather than the scientific tradition of talking over what *had been* proven, would prove perilous to the conference. It was a flaw to which he was long blind.

Thus he had no inkling that those he was inviting—or the public at large—might read his invitation differently from the way in which he meant it. His whole point was to forestall the nasty, explosive encounter such claims might be expected to produce by considering the implications and standards of evidence calmly and thoroughly in advance of the evidence. For example, he wanted his conference participants to ponder questions like: What would constitute scientifically and legally acceptable proof of a meaningful link between genetics and crime? What are the predictable pitfalls or dangers in such studies? Can a complex "phenotype," like a societally defined behavior, ever be connected securely to a genotype? And what would be the appropriate response from society if such a link were to be demonstrated?

In this regard, Wasserman thought the controversy about individuals with an XYY genotype—a debate that had flourished in the 1960s—

would be an important precedent to examine at the conference. Although the normal human condition is to have only two sex-determining chromosomes (XX in females and XY in males), there is an uncommon condition in which a male possesses two Y chromosomes, making him an XYY. This is a cytogenetic problem—one that occurs during formation of the sperm—and so is genetic without being inheritable. That is, XYY males do not father XYY male offspring, yet it is the extra Y chromosome that seems to be responsible for specific, identifiable attributes of an XYY individual. These include greater-than-average height and a tendency to acne; more controversially, XYY genotypes have been linked to mildly deficient intelligence.

But this was not all. In 1965, a team led by P. A. Jacobs found that the XYY anomaly occurred in unusual frequency among the inmates of a maximum-security prison in Scotland.[6] There, XYY's are 1 or 2 individuals in 100, versus about 1 in 1,000 in the general population. Additional studies confirmed that XYY's were ten to twenty times more common in penal populations, which seemed to demonstrate a causal link between this genetic anomaly and criminal behavior.

On the basis of these early studies, it was hypothesized that this extra Y chromosome led to a predisposition toward violent behavior as well as tallness and skin problems. Newspapers and magazines carried frightening stories about the sinister power of the XYY genotype. The fear was heightened by the bizarre claim that Richard Speck, the perpetrator of a particularly gruesome series of eight murders of student nurses in Chicago, deserved clemency because he was XYY.[7]

In direct response to the awful possibility that the hypothesis was correct, a group of Boston physicians initiated a screening program on newborn male babies. Their intent was to identify and follow any XYY's detected and to compare their criminal tendencies with those of normal male babies. The study was intended to determine whether or not the XYY genotype was linked to violent behavior. The parents of XYY children in the study were to be informed of their child's defect; it seemed immoral to ask for their cooperation and then withhold such vital information.

Critics were vociferous: what of the possible influence of self-fulfilling prophecies? And what of the pain and anguish to all involved, what of the years of needless suspicion and anxiety, if no link were found? In the end, the study was stopped. And later, extremely carefully constructed studies disproved the association of the XYY genotype with violent criminality.[8] Sadly, the "common knowledge" of such a link lingered long afterward.

The fundamental flaw in the reasoning of the early studies lies in the facile acceptance of the correlation. Correlation is not the same thing as causality. While disproportionately more criminals may be XYY, the inverse is not true: XYY's are not more likely to be criminal. It has been estimated that as many as 96 percent of XYY males are normal, law-abiding individuals. Mass chromosomal screenings and interventions would waste huge amounts of money and would stigmatize over-whelming numbers of innocent males.

With this case, the door was opened briefly for a whole new type of prejudice, based on genetics and therefore structurally linked to racism, for once a link between a specific gene and criminality (or other so-cially unaccceptable behavior) was demonstrated, it would be possible to find out which population, ethnic group, or race possessed that gene in higher freqencies. Scientific data might lend strong legitimacy to discrimination against those more likely to be genetically "flawed." Compulsory sterilization, marital prohibitions, and worse could ride into town again on the coattails of genetic essentialism.

Among the liberal scientists in the genetic and biochemical com-munities who blocked the proposed screening program was Jonathan Beckwith of Harvard, who had agreed to attend Wasserman's confer-ence. Wasserman wanted to include a roundtable debate about the XYY controversy and its lessons for a new generation of genetic research on crime.

In addition to providing the opportunity for those investigating ge-netic components in behavior to argue with their critics, Wasserman also invited neurophysiologists to present their work on the biochem-ical regulation of violent and impulsive behavior and respond to those who favored an environmental and social explanation. He also wanted the participants to explore the interactions of the adversarial legal sys-tem with the results of scientific and technological studies. How is in-formation to be used and presented? What degree of certainty is acceptable, and how will uncertainties be dealt with by judges and ju-ries? What would be the benefits and risks associated with setting up genetic databanks, similar to those now used for fingerprints and mug shots?

Finally, he hoped the conference could address issues of interven-tion, screening, deterrence, and blaming. Is it moral to test a popula-tion for a genetic predisposition toward crime? If it is moral, is it also legal and in the best interests of society? Is there a point in diagnosing and identifying those with such a genetic predisposition in the absence of any good plan for intervention—curing? Or would only discrimination and condemnation be the result? And is someone to blame for a crime if he or she has a demonstrable genetic predisposition? Does crime then assume the properties of a heritable disease rather than a conscious choice?

It promised to be a fascinating and important conference. In fact, as a means of defusing a potentially explosive research issue, the concept of discussing criteria, validity, implications, and policy—*in advance* of the research being completed or presented—was brilliant. The National Institutes of Health peer review panel thought so too; they lauded Wasserman's proposal. They wrote:

> [A] superb job of assessing the underlying scientific, legal, ethical and public policy issues and in organizing them in a thoughtful fashion. . . . This proposal addresses one of the areas of greatest concern about uses of information that come from the study of the human genome.
>
> The proposed conference is carefully organized. The greatest strength lies in the impressive array of participants. . . . the agenda is quite well developed. . . .
>
> Overall an excellent grant application for a conference on a timely and important topic. It will quite likely be considered a landmark in the field.[9]

Not only did they praise Wasserman's conference, they ranked it highly and actually recommended giving him more money than he asked for.

The panel expressed concern that more practitioners, nonacademics like judges and lawyers, should participate. But their main worry was that the results were likely to be of such importance that Wasserman had not allocated enough funding to disseminating the findings. In the proposal, Wasserman explained that he would publish an edited volume of papers from the conference and a special issue of *QQ* (now called *Report from the Institute for Philosophy and Public Policy*), the journal of Wasserman's home institution; a videotape would also be made. The panel advised adding an extra $23,200 in direct costs—an increase of almost 30 percent—for three purposes. First, they wanted to ensure that the videotape was of high quality, suitable for showing on public television or in college classrooms. Second, they wanted to offer a $500 honorarium to speakers, which would obligate them contractually to provide papers for the conference volume. Finally, they wanted to double the publication run of *QQ* so it might be more widely distributed.

Wasserman's grant was approved and funded. There were congratulations all around in the institute where Wasserman worked and he set about publicizing the conference and finalizing the program. He was, he thought, well on the way to success. Even though the extra money that the study section panel had recommended was not forthcoming from NIH, Wasserman hoped to receive a supplement to his grant or to find some other source for it. And the panel's kind words simply increased his elation.

At this point, the only sign of danger that Wasserman sensed was the chance that his combination of participants was too explosive. "When the conference was approved there was a great deal of excitement that we had actually gotten these people together," Wasserman remembers. "Part of the reason I wasn't focusing on how this conference might be perceived by the wider public was that I was so concerned that the conference might blow apart internally. I was anesthetized by the study section's remarks which had been very enthusiastic."[10]

Wasserman was both ignorant of history and innocent of intent. He had never heard of Carleton Coon, neither read *Origin of Races* nor heard the bitter outcry against it. He did not realize he was still vulnerable to accusations of racism—indeed, he never considered the possibility. The fourteen-member study section panel at NIH, which included three African-American scholars, had approved his proposal without demur. Nowhere in their five-page review was any concern expressed about the issue of race or the potential that genetic factors predisposing toward crime (if such existed) might be seen as more common in any particular race. None of the proposed participants was expected to defend the proposition that races differed in their genetic predisposition toward violence. Indeed, the word *race* did not appear in the proposal, because Wasserman did not perceive this to be a central theme of the conference, and none of the few participants of African-American ancestry raised any concerns.

"To a lot of black scholars," Wasserman admits ruefully, "it seems hard to believe that I could have organized this whole thing without having thought of race as a central theme. But I say, 'Look, no one in the study section said *what about race?*' "[11]

To Wasserman's regret, his brilliant conference was about to escape from his hands; under the influence of others, it was transmuted into another being altogether. Far from being quiet and contemplative, a leafy tree in whose shade abstract and concrete issues could be thoughtfully and coolly considered, Wasserman's conference grew nasty spines and poisonous fruit that brought pain to all who came in contact with it.

Part of the transformation occurred during the inevitable lag time between the writing of the proposal and the conference itself. Like Coon, Wasserman was overtaken by historical events over which he had no control. Some eight or nine months after he began assembling his speakers, a man totally unknown to Wasserman, Dr. Frederick Goodwin, took a stand that provided a context from which Wasserman's conference could not be divorced. The conference was doomed.

Goodwin was the head of the Alcoholism, Drug Abuse and Mental Health Administration (ADAMHA), the highest-ranking psychiatrist in the government, whose personal research focused on the neurobiology of aggression. On February 11, 1992, speaking to the National Mental Health Advisory Council, he announced that the number one priority of ADAMHA in its 1994 budget would be a new Violence Initiative. Under the aegis of the Department of Health and Human Services, the Violence Initiative would formalize a frontal assault on the violence that plagues modern American life. The Violence Initiative would commandeer the same resources, single-minded focus, and determination with which the government had attacked other devastating public health problems, such as smallpox.

Goodwin is a Caucasian but his boss at that time, Dr. Louis Sullivan, is an African-American. As a physician and the director of the Department of Health and Human Services then, Sullivan was the man most responsible for designing the Violence Initiative. Sullivan firmly advocated a public health approach to the problem of violence which was "too great to be left to the criminal justice system"; in Sullivan's words, the United States had become "the most violent country in the industrialized world."[12] He was convinced that treating violence as a health problem was entirely justified by the fact that it causes a tremendous number of deaths and injuries in the United States annually—and that these rates are skyrocketing.

In his annual report on the state of the nation's health, in June of 1992, Sullivan spelled out exactly how severe the problem of violence had become. "This increase [in homicide rates] is attributed, in large part, to a rising rate of homicide among young black men. Between 1985 and 1989," Sullivan noted gravely, "homicides were up 74 percent among young black males to reach the highest level ever."[13] Indeed, it could be argued that ignoring such a blatant cause of mortality would be a dereliction of duty—possibly even an expression of racism, a lack of concern for a segment of American society against which many are still prejudiced. As a public official charged with guarding the nation's health, as an African-American, Sullivan could not ignore this crisis.

In this first announcement, Goodwin emphasized that the plan was to focus on individual vulnerability. The idea was to look for early predictors of future violence, by studying behavioral and biological markers, and to try to establish a useful pattern of intervention once those predisposed to violence had been identified. He envisioned a target population of perhaps 100,000 inner-city youths who could be helped. These twin notions might have been controversial, but Goodwin's most serious error was to try to put the problem in perspective by referring to research on other primates.

Since Darwin's day, it has been axiomatic in evolutionary biology that humans share many biological and behavioral traits with their nonhuman primate relatives. It was precisely this point that Thomas Huxley fought to bring home to Victorians in England: we are primates, we are of nature not above nature. Today, researchers commonly look to monkeys and apes to try to deduce the attributes that humans held at an earlier stage of their evolutionary history, although conclusions based on such comparisons must always be constructed cautiously. And so Goodwin, imbued in this research tradition, uttered the words that brought a barrage of criticism down on his head:

> I say this with the realization that it might be easily misunderstand, and that is, if you look at other primates in nature—male primates in nature—you find that even with our violent society we are doing very well.
>
> If you look, for example, at male monkeys, especially in the wild, roughly half of them survive to adulthood. The other half die by violence. That is the natural way of it for males, to knock each other off and, in fact, there are some interesting evolutionary implications of that because the same hyperaggressive monkeys who kill each other are also hypersexual, so they copulate more and therefore they reproduce more to offset the fact that half of them are dying.
>
> Now, one could say that if some of the loss of social structure in this society, and particularly within high impact inner city areas, has removed some of the civilizing evolutionary things that we have built up and that maybe it isn't just a careless use of the word when people call certain areas of certain cities jungles, that

we may have gone back to what might be more natural, without all of the social controls that we have imposed upon ourselves as a civilization over thousands of years in our own evolution.[14]

Huxley had known that pointing out to people that their ancestors had been monkeys would be deeply unwelcome, and did it anyway, skillfully and carefully. Goodwin apparently didn't appreciate how much more inflammatory comparing humans to monkeys would be, and tripped on a land mine—as he soon found out.

And a further question must be raised: was Goodwin projecting information about monkeys onto blacks, or was he projecting stereotypes about African-Americans onto monkeys? Few descriptions in the primatological literature of monkey societies bear much resemblance to the blood-and-sex thumbnail sketch offered by Goodwin. And no one familiar with the complexity of primate societies would underestimate, or simply deny, the many behaviors and gestures that resolve conflicts and appease aggressors[15]—the "social controls" that Goodwin seems to imply pertain only to *human* society. Certainly Goodwin's remarks echoed old and bitterly distasteful stereotypes of African-Americans as primitive (monkeylike), violent, and oversexed.

The protests in response to Goodwin's announcement reached an extraordinary level. The Congressional Black Caucus, then an organization of 26 black legislators, asked to meet with Sullivan to discuss Goodwin's "ignorance about the use of behavioral sciences that raises serious questions about his fitness to continue serving as the administrator of the Alcohol, Drug Abuse and Mental Health Administration."[16] Other individuals and groups, like Blacks in Government (BIG) or the American Orthopsychiatric Association, with nearly 10,000 mental health professionals on its rolls, also called for his dismissal. Newspaper headlines fanned the flames of indignation, particularly within the African-American community.

Goodwin apologized publicly on February 25, at a meeting of the Mental Health Leadership Forum, for his insensitivity and for causing inadvertent offense with his remarks, but it was not enough. In the midst of a firestorm of censure, Goodwin resigned his post as head of ADAMHA. Sullivan moved him into another influential but lesser position, as director of the National Institute of Mental Health (NIMH), promising that his involvement with the Violence Initiative would be minimal.

But Goodwin's words had heightened the sensitivity of two groups of citizens, groups who would form an alliance against the Violence

Initiative. One group was the African-American community, alarmed at the prospect of becoming the target of a powerful government initiative to control or eliminate certain kinds of behavior. Some in this group coalesced around Ronald Walters, a Ph.D. who was head of the department of political science at Howard University. The other was a group of activists led by activist psychiatrist, Peter Breggin, who was deeply opposed to pharmacological interventions as treatment for psychological problems and had been leading a fight to reform psychiatry for years. To these, Goodwin's words heralded the beginning of one of the most dangerous policies they could imagine: a massive program to drug and subdue a large segment of the population. They were not surprised that the target population represented poor racial minorities. Whom else would the power structure attack?

And David Wasserman's conference—the one he thought was about genetic technology and the law—would be swept into this maelstrom of racial controversy and drowned.

A Conflict Character

CHAPTER 14

PROTESTS AIMED AT GOODWIN diminished somewhat with his seeming demotion, but concern about the Violence Initiative itself escalated through the spring and summer of 1992. As yet, Wasserman and his conference played no role in the unfolding drama.

Ronald Walters was one nucleus around which anxiety in the black community crystallized.[1] Walters is known in the black community of the area around Washington, D.C., as an able and concerned activist. He is an impressive man, with a bachelor's degree in history and government from Fisk University, and an M.A. and Ph.D. from American University in African and international studies, respectively, with a spate of articles and three books on his specialties—African-American political participation and the politics of southern Africa. He advised Jesse Jackson on his run for the presidency of the United States. Best of all, Walters is no rabble-rouser, no wild-eyed, irrationally partisan radical; he is articulate, thoughtful, and as concerned with the integration of American society into a healthy whole as he is with black rights.

When Goodwin's offensive remarks made the news, Walters began getting phone calls from people in the community. Even after Goodwin's demotion, people kept calling to express their fears that the Violence Initiative was a dangerous policy and asking Walters to "do something about it." In June of 1992, he organized an open meeting at Howard University that over 100 people attended. These were mostly what Walters calls "workaday people, not academics . . . some whose children are on Ritalin [a drug prescribed for hyperactive children], some community activists . . . simply interested citizens."[2] Out of that meeting a group was formed that took the name Committee Against the Federal Violence Initiative, which was still meeting every other week at the year's end. A similar meeting was held in New York City.

The nub of the Washington group's concern was their interpretation of the intention of the Violence Initiative. They found the plans to try to identify individuals prone to violence threatening, especially since inner-city youths—which they take to mean poor black youngsters— are the designated target. The plans for intervention mentioned by Goodwin also frighten them.

Their fear was heightened by the alarming remarks of Dr. Peter Breggin, a white activist psychiatrist with strong views that contrast sharply with his calm, articulate demeanor.[3] Breggin freely identifies himself as "a nonmainstreamer"[4] of American psychiatry, despite his mainstream credentials: a B.A. from Harvard College, an M.D. in psychiatry from Case Western Reserve Medical School, and postgraduate experience at the State University of New York in Syracuse (Upstate Medical Center) and the Massachusetts Mental Health Center. He even worked for a time at the U.S. Public Health Service at the National Institute of Mental Health (NIMH), the organization that Goodwin headed up long afterward.

He is not afraid to take a novel stance but recognizes that not all his views are shared by the majority of his professional colleagues. "I'm always picking something so much on the fringe or the frontier that it takes people a while to catch up," he says.[5] The profession's catching up—or coming to agree—with him seems unlikely on several fronts, for, in his case, both agree that they are a long way apart.

What has made Breggin controversial is his attitude toward many standard treatments for psychiatric problems. From his earliest exposure to psychiatric treatment, as a student volunteer in a mental hospital in the 1950s, Breggin found his future profession "extraordinarily abusive."[6] He was particularly horrified by electroshock therapy and insulin coma treatments—both generally considered to be beneficial to and highly effective in patients with particular disorders;[7] to Breggin, simple human contact in the form of psychosocial therapy seemed so much more humane. His career as an activist psychiatrist, publicly criticizing his profession, was jump-started in the early 1970s when he learned that lobotomy as a treatment for mental illness was coming back.

"I made up my mind to be the first person to stand up in public and

242   The Evolution of Racism

say 'No, you're not going to do this to people,' " he explains. "I had no idea what I was getting into."[8] His concern was that the patients targeted for this psychosurgery were those with hyperactivity and/or problems with aggressive behavior. Soon Breggin plunged wholeheartedly into a national campaign against psychosurgery. The effect was to generate enormous publicity and to ignite public concern. In 1971, he formed the Center for the Study of Psychiatry as a base for his activism, in which his wife, Ginger Ross Breggin, is now his partner. The center is described on its letterhead as "a nonprofit research and educational network concerned with the impact of psychiatry upon individual well-being, personal freedom, and family and community values."[9]

Breggin also attacked the use of drugs in the treatment of mental illness. Breggin argued, in two textbooks, that such drugs produce brain damage and do no good—a position that Breggin admits has had "not much"[10] effect on most of his profession, though it has increased their awareness of potential dangers. For example, Dr. Paul McHugh, chairman of the department of psychiatry at the Johns Hopkins University School of Medicine, a bastion of the American medical establishment, describes the pharmacological treatment of mental illness as "*the* major achievement that has moved us from merely caretaking severely ill people to relieving their symptoms."[11] McHugh cites the careful clinical trials and double-blind tests that established the value of these therapies beyond any doubt. He also emphasizes that these drugs, like any other medication, must be administered with due care and attention to the patient's response.

The commonest examples of damaging drugs cited by Breggin are Ritalin, often used to treat hyperactivity, and the antidepressant Prozac. However, his condemnation of the biological approach to psychiatric problems is wholesale. "I don't think *any* psychiatric condition is suitably treated with drugs," he says. "They are not the answer to human problems. I would try very, very hard to use other approaches than, say, lithium for manic-depressives. I don't think it's the way to handle their difficulties in being in charge of their moods."[12] Breggin would neither initiate drug therapy nor would he advocate its abrupt cessation.

It is a position like this that leads McHugh to think that Breggin is rejecting some of the most effective treatments available to patients with particular, serious mental illnesses.

After voicing his views forcefully on an Oprah Winfrey segment, Breggin was targeted for a heated attack led by the National Alliance for the Mentally Ill, an organization many of whose members are the parents of mental patients. Their aim was to have Breggin's medical license revoked because—they claimed, inaccurately—he had dispensed medical advice over the air by telling mental patients to stop taking their medications. Using a tape of the program and a transcript of his remarks, Breggin counterargued successfully that he had cautioned against abrupt cessation of psychiatric drugs, which can be dangerous; that he had been expressing a scientific opinion, not prescribing medical treatment; and that his remarks were protected by the First Amendment. Breggin's exoneration was complete: the false charges were expunged from the record, and Breggin received an apology from the medical board. Breggin believes it was as a direct result of the publicity surrounding this situation that he was at last able to find a publisher for *Toxic Psychiatry*, a mass market book on his views, in 1991.[13]

Breggin had long-standing convictions about the misuse of drugs to treat psychiatric problems. When he heard about the Violence Initiative he perceived its real intent to be drugging and sedating black youths. To Breggin, what Goodwin or Sullivan said was irrelevant, because their cloaked intent was obvious. He says:

> There could never be any doubt that the proposed "intervention" was pharmacological, because that's what Fred [Goodwin] knows. This is what he does [for his own research]. He has systematically purged NIMH of all psychosocial research. There couldn't be anything else other than drugs, shock treatment, or incarceration; they don't promote anything else; it was a foregone conclusion.[14]

Breggin appeared on Black Entertainment Television, speaking against the Violence Initiative, and spoke at Ronald Walters's community meeting at Howard. Breggin is a master of the sound-bite that etches itself in his listeners' minds; his remarks on this topic were no exception.

"It [the Violence Initiative] is the most terrifying, most racist, most hideous thing imaginable," he says, in a turn of phrase that has gotten him quoted by many newspapers and magazines. "It's the kind of plan one would associate with Nazi Germany."[15]

The comparison has been made by others as well, who see a sinister intent in the casting of a social issue as a public health problem of a racially identifiable group—precisely the strategy Hitler used. Breggin also spoke to the Congressional Black Caucus, spelling out the ominous implications he draws from the Violence Initiative:

> The idea of looking for genetic or biological markers for violence in inner-city children is scientifically unsound.

Meanwhile, more than a million children in America are being drugged with Ritalin to make them more docile in school and at home. This, too, is being done without sound scientific evidence and without an appreciation of the harmful results. Inner-city schools already play a major role in pushing and sometimes coercing families to take Ritalin, a drug pharmacologically related to amphetamines and cocaine. The Violence Initiative will result in still more massive drugging of inner-city children, not only with Ritalin, but we predict with Prozac, Zoloft, and other new drugs that affect the serotonergic neurotransmitter system of the brain.

While not specifically discussing drugs, Goodwin focuses on the need to correct presumed imbalances in the serotonergic neurotransmitter system. . . . Drugs are the only possible cheap, effective intervention into the lives of tens of thousands of children. . . .

The federal government's Violence Initiative is the ultimate extreme of blaming little black children for the problems of society, such as racism, poverty, hunger, inadequate or absent health care, the decline of the schools, unemployment, police brutality, a destructive welfare system, and despair over the future.

It is Orwellian for the federal government to plan to identify little children [predisposed to violence] in great numbers through the school system for early preventive treatment.[16]

Breggin calls his role in these controversies one of "publicity and education." He adds, "We do public education and hope a power broker will follow up on it."[17]

Walters and the Committee Against the Federal Violence Initiative followed up. Walters (on behalf of the committee) began writing letters: rational but powerful letters expressing the concerns and fears of the group he represented, letters directed to members of Congress and to Sullivan.

Though Breggin's charges attracted considerable media attention, the officials in charge of the Violence Initiative disagree with his view of the intentions of the program. Goodwin flatly denies any plans for mass biological intervention and has sketched plans for counseling and special school programs: "It's basically a super Head Start program," according to authorities at NIMH. To them, Breggin's suspicions of a conspiracy to sedate black youths amounts to "a nefarious twisting of what we're doing" that was "pieced together [from Goodwin's remarks and Wasserman's conference] to form a picture of a grand scheme that doesn't exist."[18]

Under attack, Sullivan stood firm in defending the Violence Initiative.

"I will not apologize for doing what I can to reduce the number of young lives that are being tragically destroyed by violence," he said. Sullivan accused those who call the initiative a plan for genocide or a new holocaust aimed at blacks (including Breggin) of leveling "false and inflammatory accusations" that have "grossly and irresponsibly mischaracterized" the work of the Department of Health and Human Services.[19] He emphasized that the great majority of the work on violence focuses on social and environmental factors, with less than 1 percent of the Department of Health and Human Services' funds going to research exploring genetic links. Even Goodwin, who to his critics personifies the genetic explanation for crime, spoke at length of "environmental poisons"[20] responsible for criminal behavior in an address to the American Psychiatric Association in May of 1992. He enumerated such factors as sexual or physical abuse by parents, desensitization to violence through constant exposure, parental disharmony or divorce, poor schools, and alcohol or other substance abuse that have been shown by research to be linked to crime.

But Goodwin and Sullivan were making an argument that may be too subtle for critics anxious to identify simple black-and-white dichotomies, like nature versus nurture, in identifying causes of criminal behavior. Single-factor explanations have long been proven insufficient, Goodwin says: "These days to talk about biology versus behavior is anachronistic. Biology versus psychosocial forces is anachronistic. The question is, how do psychosocial forces and biological factors interact with each other and can we find experimental ways to tease apart the relative contributions?"[21] In other words, Goodwin believes that nature interacts with nurture, that behavior is conditioned by both biological and social factors, which must be disentangled if they are to be understood. But this rational approach cannot be taken without first admitting that there may be some genetic or biological component to crime, and many of Goodwin's critics are emphatically unwilling to concede this point.

The molten criticism of the initiative continued to flow, in words ranging from the critical to the inflammatory. A headline reports "Study to quell violence is racist, critics charge,"[22] with an accompanying photograph of African-American activist Cindy Owens; "Genetic Screening—The NIMH's Blueprint for Genocide"[23] and "Urban

Quackery"[24] scream the front pages of other publications. "Natural Criminals?"[25] queries the cover of a Black Muslim newspaper in bold, black type that is interspersed with images of young black men, some holding their fists aloft in Black Power salutes.

Through the spring and early summer of 1992, Wasserman watched this mounting furor with academic interest, from the sidelines. "I thought this would animate my conference," he remembers wryly. "I wasn't happy about Goodwin's remarks, but they made the issues more real, more vivid; I thought the controversy would make the conference more interesting."[26] And he proceeded with his scholarly activities, anticipating his conference in October. He was for a time preoccupied with editing an issue of the newsletter of the Institute for Philosophy and Public Policy on the subject of racism and discrimination; ironically, it displayed an advertisement for the conference on its back cover. At the same time Wasserman needed to design the brochure for the conference, not anticipating that this flyer would be deadly to his plans.

At the top of the brochure, he placed the University of Maryland seal, on the left, and the logo from the National Center for the Human Genome Project, obtained from the program officer administering the grant, on the right. Over them was the conference title, **Genetic Factors in Crime: Findings, Uses, and Implications**, in boldface, the dates, and the words "A conference sponsored by The Institute for Philosophy and Public Policy and The National Institutes of Health."

What Wasserman calls the "fatal paragraph"[27] was the one that opened the brochure:

> Researchers have already begun to study the genetic regulation of violent and impulsive behavior and to search for genetic markers associated with criminal conduct. Their work is motivated in part by the early successes of research on the genetics of behavioral and psychiatric conditions like alcoholism and schizophrenia. But genetic research also gains impetus from the apparent failure of environmental approaches to crime—deterrence, diversion, and rehabilitation—to affect the dramatic increases in crime, especially violent crime, that this country has experienced over the past 30 years. Genetic research holds out the prospect of identifying individuals who may be predisposed to certain kinds of criminal conduct, of isolating environmental features which trigger those predispositions, and of treating some predispositions with drugs and intrusive therapies.[28]

It was taken verbatim from his proposal, contrary to charges leveled at him later, with a single trivial change. Two words—"and neurobiological"—were deleted from the beginning of the last sentence in the proposal, where they followed the first word.

"I did what seemed the most conservative thing," he recalls. "I took the title of the proposal and the introductory text from the proposal [to explain what the conference was all about]. It was the same thing that judges do when they repeat language confirmed from an appellate decision."[29] He also borrowed some of the language from the study section's summary of the proposal, describing the types of researchers who would participate (they were also listed by name on the reverse of the brochure). Finally, he condensed the description of the goals of the conference presented in the proposal, making clear that different points of view were expected and even welcomed. But clarifying and narrowing the areas of disagreement were reasonable hopes and Wasserman gave examples of some anticipated debating points:

E.g., whether specific findings are replicable [or valid]; about the interpretation of that research, e.g., whether accepted findings are generalizable to other populations or other criminal offenses; and, finally, about moral and policy issues raised by the research, e.g., how does the capacity to predict and explain misconduct affect the appropriateness of blame and punishment?[30]

Because Wasserman had no idea of the depth of the fear that young black males might be drugged, he included two sinister questions that described the topic of the final session (Intervention and Treatment): "Can drug therapy ever be benign? To whom should it be offered, on whom should it be imposed?"[31] With these words, he administered a fatal dose of poison to his conference. A tentative schedule and registration information completed the brochure.

The brochure was mailed to about ten thousand professionals who might be interested in the conference. Although he was aware of the barrage of criticism leveled at Goodwin and Sullivan, Wasserman had no idea that the language of his brochure was superbly suited to provoke a violent response from Walters, the committee, and Breggin, who was sent a copy of the brochure by "someone at the University of Maryland who wishes to remain anonymous."[32]

"I never considered that we'd be linked with the Violence Initiative," Wasserman admits.[33]

Once Walters and Breggin received the brochure, neither needed further proof that Wasserman's conference was part and parcel of the same plan that fueled the Violence Initiative. While Breggin scrupulously avoids the word "conspiracy," he sees a deliberate and sinister pattern in American history, from its beginning up to the present time.

> I would never speak of conspiracy [he says]; it is a way of trying to denigrate or dismiss the reality that people do get together and plan and organize. Ginger [my wife] thinks "conspiracy" was invented to discredit anyone trying to criticize the establishment. Of course they [the government] are consciously planning things, the same as we are. Accusing your critics of being paranoid is the personal equivalent of labelling the criticisms of the establishment as conspiracy theory.
>
> Is there a planned conspiracy to destroy the black community? I don't know, but that has been the effect of public and government policies since slavery. They have stripped blacks of all of their power and identity systematically. There is a long, long history of suppressing black people, so the real question is: if there's no real conspiracy, tell me, when did it end? Not with slavery, not with reconstruction. . . . This country has been organized around the suppression of black people for a very long time.[34]

Walters tends to discredit the notion of an intent to harm, but "if you're an African-American you have enough reasons to be suspicious," he says, citing the infamous syphilis study carried out at Tuskegee, in which infected black men were denied treatment, as precedent for the harmful effects that government-sponsored medical interventions have had on the black community.

"We didn't make this up," he continues defensively. "There was no doubt about whom Goodwin's Violence Initiative was going to target." And he finds it "curious" to characterize violence as a public health problem. "Violence is a public health problem only to the extent that people die from it," he explains. "But it is not a public health problem in terms of the origin of the problem."[35]

For Walters, the causes of violence are socioeconomic, not biological. In a paper delivered in 1993 to the American Association for the Advancement of Science, Walters paints the rise in urban violence in the 1980s and 1990s as being "fueled by economic deprivation"[36] and observes that "racism and deprivation prevented black upward mobility in virtually every major area of life."[37] He also draws a causal connection between public policy and the rise of violence.

> What we know is that there has been a massive withdrawal of so-
> cial resources, especially in the past 12 years during the Reagan
> and Bush administrations, causing poverty to deepen [Walters
> writes]. It is not accidental that this period also witnessed the rise
> of the drug trade in most major inner city areas, and the collater-
> al rise of violence and murders as well.[38]

Walters is troubled by the public health model of violence—the paradigm of violence as a disease—which

> shifts the focus from the social status of groups to individual mal-
> ady. . . . Even if some level of violence is a part of the cultural life
> of African Americans, the best intervention is the availability and
> provision of productive life choices which would give individu-
> als a valid alternative to violent or criminal behavior.[39]

In contrast, the aim of the Violence Initiative is "patently . . . social
control"[40] and possibly even coercion, two tactics feared, with good
reason, by the black community.

Into this climate of suspicion and fear came Wasserman's conference
brochure. The brochure gave an awful plausibility to Breggin's claims
that the government was legitimizing the idea of a genetic component
to crime. Breggin cast the conference as "the tip of a much larger and
more dangerous iceberg, the Federal 'violence initiative,' "[41] a deliberate preparation of the minds of the public to accept the wholesale sedation of black youth. Similarly, Walters felt the conference could not
be divorced from the Violence Initiative, despite Wasserman's protes-
tations to the contrary.

"We didn't necessarily conceive of Wasserman as the bogeyman,"
Walters says. "I thought he was purely clumsy in the way he put the
conference together; either that, or disingenuous. . . . Minority representation [in the conference] was scarce to non-existent and this didn't
give you much confidence that they would be able to handle the subject of minorities with much sensitivity."

Walters continues with his criticisms:

> I said [to Wasserman], "You have to see this against the backdrop
> of the Violence Initiative; some are advocating using drugs on
> inner-city kids."

He said, "But we're not going to be discussing race."

I thought that was an extremely naive rejoinder from someone putting together a conference on this explosive subject. He was not in command of the issues and the attendant concerns. You cannot discuss the science without discussing the social causes of the increase in crime; you cannot discuss it without discussing treatment, without talking about *who* you're going to treat and *how* you're going to treat. . . .

If you [Wasserman] don't believe that there is a subtext on the causes of crime in the inner city driving the conference, what confidence can *we* have that you will be able to handle the issue with competence?[42]

The first time Wasserman heard that his conference was now about race, not about genetic technologies and the law, was while he was on his honeymoon in Portugal. He received a phone call saying there was a problem with the conference raised by Breggin and Walters's group. Wasserman was dumbfounded. He suggested that anyone who complained be sent a copy of the proposal, so they could see who was participating and would realize that the conference was neither focused on the genetics of any particular race nor stacked with those who endorsed a genetic causality for crime.

After Wasserman's return to the United States, Breggin welded the conference to the Violence Initiative with hot words reported in the media. Walters, for the Committee Against the Federal Violence Initiative, wrote a letter of protest to Dr. Bernadine Healey, then the Director of the National Institutes of Health—the superagency of which the Human Genome Project was a part. Criticisms were voiced on radio talk shows and in print. Sullivan, in an attempt to deescalate the controversy, formed a blue-ribbon panel to monitor the Violence Initiative and coerced an unwilling Walters into participating. The panel requested a copy of the ethical guidelines for the Human Genome Project and found, to their dismay, that there was none, which did not enhance their confidence that Wasserman's conference was being carefully overseen.

At this point, the program director at the National Center for the Human Genome Project, Eric Juengst, reassured Wasserman; in his experience, there were often flurries of concern about conferences that usually faded away. He suggested issuing a carefully worded press release explaining that the conference was not *endorsing* the existence of a link between genetics and crime but was rather *examining*, skeptically, the claims for such a link. But there was no stopping Breggin's visible crusade or Walters's powerful protests. Various chapters of the NAACP (National Association for the Advancement of Colored People), who had been receiving phone calls from the public, added their influential voices to the chorus of objections.

The strangest incident in the entire drama[43] occurred one evening. According to Wasserman, he, his new wife, and Alan Strudler, a colleague of Wasserman, were sitting in a pizzeria in Bethesda, discussing strategies for dealing with a "diatribe" about the conference that Breggin had printed in the newsletter of his Center for the Study of Psychiatry. Suddenly, Wasserman was haunted by a notion he describes as Dickensian that Breggin would be there in the restaurant. He mentioned his fear; Strudler, playfully, said, "Let's find out." He stood up and called, in a clear voice, "Peter Breggin?"

The man sitting behind them looked up and said, "Yes?" Wasserman turned to face his nemesis, asking if the woman sitting with him was Ginger Ross Breggin, his wife and the main organizer of Breggin's protests against the conference. She was. "And *who*," intoned Breggin, "are you?"

"I'm the guy whose conference you've been slandering," replied Wasserman.

They talked for a while, Wasserman too shaken and stunned to eat. Wasserman explained the origin of the conference and its innocent intent, stammering in his earnestness and embarrassment. Breggin explained his perspective: that there was a deliberate government policy aimed at young black males that was essentially of genocidal intent. Any activity lending respectability to such a design was complicit in his view. While Wasserman agreed with Breggin that the social policies of the Bush administration were abhorrent and had worsened the situation of urban blacks, he saw no evidence of intentional malice.

"I didn't feel unsympathetic to the impulse to find something deliberate and concerted in that," Wasserman explains, recounting his conversation with Breggin in the pizzeria, "but I still didn't believe that there *was*. I just don't think people are capable of those kinds of feats of social engineering."[44] Besides, he argued, the conference had had its genesis among the left-wing critics of links between genetics and crime, like Paul Billings, a medical geneticist and collaborator of Jonathan Beckwith. How could it be part of a plan to prove crime was genetically caused?

To help Wasserman understand the sensitivity blacks had to his conference, Breggin proposed a metaphor—not the "Nazi Germany" comparison that was featured so prominently in news accounts of the dispute, but another. Speaking as a Jew, to a Jew, Breggin asked how Wasserman would feel about a conference on genetic factors in junk bond trading. Wouldn't the B'nai B'rith rightfully shut such a conference down? Wouldn't every Jew in America be outraged?

Yes, Wasserman agreed, but because the idea that there might be genetic factors for something so specific as insider trading was absurd and ludicrous, even if Jews were often involved. He thought that the idea that there might be genetic factors in crime in general was *not* absurd and ludicrous. Besides, the focus of Breggin's hypothetical conference was clearly on Jews, whereas Wasserman's conference was not focused on blacks. They parted with somewhat greater sympathy for one another but with no softening of their diametrically opposed views.

Breggin's understanding of Wasserman's conference seemed as resistant to modification as his perception of the Violence Initiative; the words of those who planned them made no impact. *However* the conference had started, it was being used as a pawn, as a prelude to the government's real agenda; *however* it proceeded, and regardless of who its participants were, its mere existence promoted a racist view. "The title itself gives the whole thing away," Breggin remarked later, once again reading a hidden intent behind innocent words. "The subject itself [genetic factors in crime] doesn't exist and it has always had racial implications and little more than that."[45]

In July, the flames of public concern burned deep into Wasserman's conference. The problem passed out of the hands of Eric Juengst, the program director in charge of Wasserman's grant, and up the administrative ladder. Bernadine Healey, the director of NIH, and John Diggs, her deputy director for extramural research, froze the funding for Wasserman's grant until "further discussions" occurred "concerning the implications of the conference that is planned." They asked Wasserman to engage an advisory committee to "review all conference materials and plans with the goal of identifying modifications that will be needed to alleviate the concerns or clarify the issues that have been raised." Those concerns were "associated with the sensitivity and validity of the proposed conference."[46]

To anyone familiar with the federal granting system, it was shocking that Healey had intervened directly. A cardiologist with impeccable medical training and considerable high-level administrative and policy experience, Healey is an attractive blond woman with an incisive intelligence. She projects that unshakable confidence in her decisions which surgeons often display, a trait some colleagues find disconcerting and others intimidating. Healey took on the job of heading up NIH after several men had turned down the position and proceeded to make her mark by various bold decisions, several controversial. She is not one to back down easily from a stand once taken. Diggs, her deputy director of extramural research, is an African-American with a Ph.D. in physiology and years of administrative experience.

When the protests began reaching her office, Healey apparently contacted some members of the study section that had approved the conference proposal, asking them to defend their decision. They reread the proposal and refused to back down or downgrade their ranking of the grant. She went ahead with the freeze anyway, feeling it was the only responsible reaction to the protests.

By the time the freeze was imposed in midsummer, Wasserman was already trying to respond to the public outcry—and to save his conference—in various ways. He had told his program director, Juengst, that he was more than willing to convene a panel including representatives from the black community and others and to listen to their suggestions. He had already agreed to change the title of the conference to something less offensive than "Genetic Factors in Crime: Findings, Uses, and Implications" and to modify the brochure. (In fact, Wasserman had originally altered the title from "The Role of Genetic Research and Technology in Predicting and Explaining Criminal Behavior" in order to squeeze it into the space on the form issued by NIH for proposal cover sheets.)

Wasserman also offered to organize one or two new sessions to deal explicitly with questions of race—though he notes that the idea did not come from, for example, Troy Duster, a black sociologist who had agreed to participate and who had reviewed five drafts of the brochure, but from a white, Jewish scholar in Minnesota. Wasserman is defensive that he was not alone in his failure to predict how the African-American community might perceive the conference. Wasserman even proposed to do away with the conference and simply compile an anthology of papers on the subject.

The freezing of funds on Healey's direct authorization shifted the grounds of the argument dramatically and her questioning the study section outraged Wasserman: "Study sections are like juries," he observes, using a metaphor drawn from his legal background, "they're *sacrosanct*."[47] He is as prone to see events through the prism of his legal background as Walters is to view them from the perspective of black political concerns or Breggin is to place them in the context of sedative psychiatry.

It was now, in Wasserman's eyes, an issue of law and academic free-dom. The whole context of the conference and its problems had been shifted to new ground. Wasserman perceived Healey's actions as a dangerous circumvention of the entire peer review process. Other sci-entists agreed, like Bill Bailey of the American Psychological Asso-ciation, who was quoted as referring to the freeze as undermining the integrity of the entire peer review process[48]—a reading of the situa-tion hotly contested by NIH spokesmen. Wasserman and Victor Med-ina, the director of sponsored programs at the University of Maryland, were certain that the freeze was baldly illegal because it unilaterally imposed new conditions on a refereed award that had already been granted, with the terms and conditions fully spelled out at the time.

What followed was a series of skirmishes conducted by letter and through the media. Medina, on behalf of the University and Wasser-man, fenced with Diggs and Healey of NIH; Walters and Breggin, as representatives of the public, continued to express deep concern about the proposed conference; the media reported a series of statements from all sides that showed how sharply their perceptions differed.

Medina's position was that Wasserman simply could not and would not discuss changes or modifications of the conference if these were a new condition of the grant because such new conditions were illegal. Walters continued to monitor the situation and, less than a month after Medina sent his letter *protesting* the freeze, Walters wrote Healey on behalf of the Committee Against the Federal Violence Initiative *prais-ing* it and urging a complete cancellation of the conference on the grounds that the

> sloppy and insensitive handling of this subject, even with the best of intentions, could result in helping to further legitimize a line of thinking about social problems which sees biomedical inter-vention as the major corrective tool rather than the rearrangement of social resources in order to provide valid alternative life chances to crime. . . . Thus it is only rational that we view this conference within the perspective of the so-called "violence ini-tiative." . . . Regardless of protestations to the contrary that these are "separate" projects and only incidentally related, to us, they are most certainly connected.
>
> We, therefore, see this conference, as well as the activities and genocidal attitude of Dr. Richard [*sic*] Goodwin and his col-leagues, as a clear and present danger to the African American community. . . . We only ask that your office exercises the kind

of sensitivity to our request to withdraw funding from the conference altogether. . . . We will continue to be very interested in other aspects of the so-called "violence initiative."[49]

Diggs, speaking for NIH, replied to Medina, explaining that further discussions were needed "due to unanticipated sensitivity and validity issues raised following publicity about the conference." The justification for such an action was to "assure and protect the interests of the public and the proper use of grant funds." Diggs also objected to the claim for NIH sponsorship on the brochure. While it is a condition of such awards that federal funding sources be acknowledged, Diggs felt "the designation of NIH as a joint sponsor is inappropriate in that it implies our direct involvement in the planning and development of the program."[50]

Media coverage of the controversy now took on a new flavor. Whereas the first round of articles had emphasized charges of racism and Nazi-like programs, examining the perceived link between Wasserman's conference and the Violence Initiative, the concern was now about the appropriateness and legality of canceling the conference. Wasserman and the University of Maryland stonewalled NIH, refusing to admit the legitimacy of the freeze.

Perhaps in response to the increasingly negative tone of the publicity about the freeze, Diggs wrote again more strongly. He quoted passages from the introductory paragraph—the one lifted without any meaningful amendment from the proposal—claiming that "These statements not only inflamed public opinion but also represent a radical divergence from the topics for which the grant was awarded." Wasserman's statements to the press (that Healey's actions were undermining the peer review system) were called "disingenuous" and "intellectually dishonest" by Diggs. NIH was responsible for expending "public funds in a socially accountable manner" and Healey was simply exercising "proper public stewardship" in responding to expressions of "extreme dismay about the conference brochure" which were labeled in one letter of protest as "dangerous."

Finally, Diggs laid the responsibility for the difficulties squarely in Wasserman's lap.

In structuring and certainly in advertising the conference, Mr. Wasserman grotesquely distorted the nature and scope of the meeting that was originally approved through the peer review process. It was the conference brochure, designed and disseminated outside of that process, which generated this controversy and resulted in the scrutiny now directed at the conference.[51]

Factually, Diggs was in error in his criticisms of the conference brochure. The wording was so nearly identical with that of the proposal that it cannot fairly be termed a radical divergence or a grotesque distortion; too, sponsorship is a standard term for acknowledging financial support.

But in another sense, Diggs identified a telling point: the brochure *reads* differently from the proposal. It is shorter, so some of the emphasis on challenging the genetic claims is softened or omitted. The fundamental purpose of the conference—the thinking through potential problems and uses of future claims for genetic links to crime—is obscured. Sentences that, in the context of the full proposal, are not alarming stand starkly on their own in the brochure and assume a more troubling import.

Those who, like Wasserman, see the brochure as the offspring of the proposal must strain to understand its reading by those who read only the brochure (or read it first) in the context of the Violence Initiative. As a brochure about a conference evaluating evidence for genetic factors in crime and seeking to develop intelligent and responsible reactions to such claims, the words can offend no one; as a brochure about black predispositions to violence, they must offend many.

October 9, 1992, came and went without a conference on genetic factors in crime being convened. On April 16, 1993, NIH terminated the grant altogether. Healey's statement on the subject was: "We as scientists and physicians believe it was an inappropriately ill-conceived, dangerously inflammatory conference in the way it was promoted. It would have been socially irresponsible for it to go forward."[52]

But that was not the end of the saga, for Wasserman and the University of Maryland appealed the termination to the NIH Grant Appeals Board. On September 3, 1993, the board ruled 7–2 in Wasserman's favor, noting that the charges of distortion and misrepresentation were "not substantiated" and that NIH was "unreasonable" and "erroneous" in failing to work with him on revising the brochure. However, with a wonderful stroke of insight into the paradoxical nature of the situation, the board also observed that, had NIH let the conference occur on schedule, it might have been perceived as "an affront to the black community." The decisions directs NIH and the University of Maryland to work together to study the feasibility of resurrecting the conference and to agree on a plan that must be approved by the National Center for Human Genome Research.[53] It remains to be seen if a civil agreement about such a conference can be reached.

As for the Violence Initiative, NIH is now reviewing all of its research on violence on both ethical and scientific grounds. After holding both public and closed sessions on the issue, NIH has already reapproved most of the 284 current studies; none has been rejected definitively. Suggestions from the review committee for future NIH actions on violence include doubling funds for studying violence prevention and for assisting victims; establishing a monitoring group to "ensure that violence research does not harm individuals or sensitive communities"; funding more long-term and social research on violence; and increasing training programs (especially for minority scientists) in violence research.[54]

The public debate—the planning, the cool-headed assessment, the setting up of standards of proof in advance—is still desperately needed, and still lacking. If the controversy demonstrates anything, it is the importance of holding a thoughtful, rational, and public discussion of these issues. But it has also raised substantial barriers to such a discussion ever taking place.

Like Coon's book, Wasserman's grand project was damned as much for what he did not say as for what he did. Wasserman failed to dissociate himself from the Violence Initiative, or from potentially racist charges that there were links between genes and crime, because he saw them as unrelated to his conference. The very absence of civil and scientific discussion of race and racial matters—the tacit embargo on the subject imposed by the scandals of the 1960s—produced a climate of suspicion, accusation, and distrust in which two utterly divergent interpretations of the same events could flourish. Wasserman could believe a discussion of genetic factors in crime did not need to focus on racial problems as a central theme, and Walters and Breggin could perceive little else but the direct danger to the black community in the same discussion. Because they are all intelligent, well-meaning men, they could talk their differences over together, but they are simply unable to reconcile their views.

The public consciousness is inflamed to such a raw and painful state that it is almost impossible to discuss the issues rationally. This problem troubles Walters, as a representative of the African-American community and a political scientist. To him, the conference is not the main concern, however precious it was to Wasserman. "The conference was a sort of a footnote," he admits. "It was not at all important in my mind. The far more serious issue is the Violence Initiative."[55] And yet he concedes that the conference worried him.

> I would want to know why such a conference on genetic factors on crime should be held; what good could come from it? If what he wanted was to talk about the implications, the standards of proof, and so on, there should have been a far different agenda and different participants. . . . You couldn't get there with the way he had structured the conference; it couldn't arise from a public confrontation between dissenting views about the validity of the scientific issues.[56]

And Walters explores an argument structurally similar to those made before, by Virchow and by Washburn and Dobzhansky. He explains:

> I would argue that there are types of research that shouldn't be done, and the grounds on which I argue are civility. You have to decide what issues are going to breed such tensions in society that there would be no return back onto the path of civility. If no group takes the responsibility for that [decision], then everything goes and you're headed down a road that is irretrievable in terms of the notion of people living together in a civil society. . . .
>
> If you say to me that I am predisposed to violence, you have further deepened the chasm between us. A cost/benefit rationale must come into play here: do you bind society up more by ignoring that kind of research or do you contribute by finding it? If you find it, then logically you must intervene to try to cure it. Well, you're not coming after *my* kids!
>
> [Some research] breeds into society a conflict character that I understand as a political scientist. . . . There are types of research that shouldn't be done because the danger to society is so great. We are on the precipice of something very important and very dangerous.[57]

The irony is that both Walters and Wasserman feel that preserving civility and preventing the fragmentation of society into mutually distrustful factions is important. In this, they are surely right, for there can be no discourse, there can be no progress or planning or strength in a society in which there is no sense of unity binding its members. The restoration of civility and good faith, the damping of the voices of fear and hatred, must take precedence over anything else.

Where they disagree is in their assessment of the role of Wasserman's proposed conference. Wasserman's vision was of a conference that might produce wisdom, that might yield conclusions which would vaccinate society against painful struggles and chasm-producing accusations yet to come. "Our intent was to raise the level of public discussion of these issues," says Wasserman sadly. "At least we've raised the volume."[58] He predicts that the claims for genetic factors in crime will be made, even if his conference is defunct.

Walters agrees that such claims are highly likely to be made and will most probably be used against the black community in America. He would seek to rob the claimants of their scientific justification by refusing to fund research that links genetics and crime or violence. The flaw in this argument is that lack of knowledge will not prevent the claim being made; it will only inhibit the scrutiny of the claim's scientific validity and implications. In Walters's view, the worst possible outcome of such research would be that a genetic predisposition to crime or to violent behavior might be established.

This hypothetical outcome must be thought through to its broader implications. Here, Stephen J. Gould's discussion of the heritability of a similarly controversial trait, intelligence, is pertinent. He writes: "The hereditarian fallacy is not the simple claim that IQ is to some degree 'heritable.' I have no doubt that it is, though the degree has clearly been exaggerated by the most avid hereditarians. It is hard to find any broad aspect of human performance or anatomy that has no heritable component at all."[59] Substitute "criminality" for "IQ" in that sentence, and it remains true.

But, as Gould points out, a trait's heritability is not the same as the inevitability of its expression; the fact of heritability says nothing about the possibility of environmental modification of the expression of the trait. The same point applies to violence (or the converse, a tendency to resolve conflicts peaceably) if it is also heritable.

Gould uses the example of height, a trait that has a higher heritability than has ever been proposed for IQ, criminality, or violence, to make his point. Assume that tall, prosperous people living in America and short, destitute people living in the Third World share a 95 percent heritability for height; this would mean that relatively tall parents will have tall offspring and relatively short parents will have short offspring with a very high degree of predictability. However, despite this high degree of heritability, if adequate nutrition and medical care were provided to the Third World population, the average height of the next generation would be virtually guaranteed to increase. So, too, might a genetic tendency to violence (if such exists) be suppressed by environmental changes and a tendency toward peaceable, cooperative behavior be enhanced.

Estimates of the heritability of behavioral traits vary, but are instructive nonetheless. Intelligence is accorded heritability at any figure between 20 and 80 percent; violence or criminality is placed at well below 50 percent. Of course, both criminality and intelligence are artificial constructs, composed of many different tendencies and abilities. Representing either as a single entity or number involves serious distortion of reality in the cause of simplicity.

What these heritability estimates denote is a subtle but important point. In 1969, Arthur Jensen claimed that intelligence has a heritability of 80 percent.[60] In other words, in the particular population that he studied, 80 percent of the *variance* of IQ values about the mean (defined as a score of 100 in IQ tests) is explained statistically as a function of genetic factors such as the IQ values of the parents of that population. The remaining 20 percent is thus attributed to nongenetic factors, such as environment, socioeconomic status, cultural value placed on learning, or quality of education. The point is that a heritability of 80 percent is an estimate based on the data on a particular population; this figure does not mean that 80 percent of a particular individual's IQ is determined by his or her parents' IQs.[61]

And these numbers give a false sense of precision. Not surprisingly, different studies using different measures of intelligence or other behavioral traits yield different estimates of heritability. More to the point, different studies using the same measure of intelligence (or criminality) on different populations also yield different estimates of heritability.

To some, these figures are a source of horror, suggesting that intelligence, criminality, and other aspects of behavior are to a large extent genetically preordained. But these heritability estimates are also a source of great hope. If the hereditarians are correct, then a very substantial percentage of intelligence or criminality is amenable to environmental modification. Consider what celebrations would occur if violence were diminished, or intelligence improved, in a population by even 15 percent! (Ironically, even if performance on intelligence tests were improved, the mean IQ score would remain at 100, by definition. What would change is either the distribution of scores around the mean or the number of questions that have to be answered correctly in order to achieve a score of 100.)

The trajectory begun with Darwin has run its course. No one has sought to provoke a bitter controversy, but the value of the differences among humans has reached out its sticky pseudopods and engulfed the unwary over and over again. The monster cannnot be outrun; it threatens us all. There is real danger here, as Walters well knows. To date, we have feared to wrestle with it openly; we have turned our heads and shielded our eyes from the horror of the problem. Rather than face the monster, we have played, instead, at politicizing first evolutionary theory and then genetics, for we are intrinsically political animals and it is a game that comes naturally. We have fought each other—called each other names, accused each other of sinister intent, promulgated bitter insinuations—instead of fighting ignorance. In so doing, we have given the hate-mongers time to feed the monster. It has swelled on a steady diet of racial divisiveness, lies, and half-truths until it is strong enough now to destroy us all.

The question is now: can we prepare ourselves for this new level of debate about the value of the differences, or will we succumb, once again, unwary?

Epilogue: Valuing the Differences

THE STRUGGLE TO DETERMINE the value of the differences among us persists.

The questions recycle again, always pungent, never resolved. What *are* the differences among us and what is the greater significance of this diversity? Who is responsible for the misuse of innocent information? Are there topics that cannot, *should* not be discussed because they are such tempting fuel for wicked fires? Who decides what is too dangerous, where the lines must be drawn? How can we know where the truth lies?

But there is an issue deeper than these in this long argument that has stretched from Wilberforce to Wasserman, from Huxley to Healey: Are we to discover *who we are* and *of what humanity exists*? May we look unabashedly in the mirror? Have we the courage and the intelligence to face the truth about ourselves?

We must. Ignorance is never a solution.

People are different from one another and in this sense human races exist. But human groups evolve, have evolved, will evolve; with each birth and each death the genetic attributes of human populations alter. Drawing a line around an ephemeral entity like a human race is an exercise in futility and idiocy. The apartheid former government of South Africa foundered in its attempts to define racial categories and created a system rife with inconsistencies.[1] Even Carleton Coon, arch-classifier of humans that he was, acknowledged the impossibility of racial classification: "Not every person in the world can be tapped on the shoulder and told: 'You belong to such and such a race.' "[2]

The only way in which human races could be truly pure or readily separable would be if they were reproductively isolated—and yet everywhere that two groups of humans have come together, they have interbred enthusiastically. Still, there is undeniable variation among humans, both physical and mental. While we do not all possess the same abilities, no more than we are all the same height or weight, collectively we possess a dazzling array of talents, for song, for joy, for building, for destroying. This diversity is our greatest gift and poses our greatest challenge.

Those attributes that have been so inflammatory, criminality and intelligence, may well also have a genetic component. But having a genetic *component* is a very different thing from being under genetic *control*. It would contradict all of our knowledge of modern evolutionary biology if a complex, behavioral feature such as intelligence or criminality were purely genetic. There is surely a tangled web of factors—genetic, social, economic, and environmental—that run from the genotype to these aspects of our phenotype. Our only hope lies in the certainty that these attributes are subject to tremendous environmental modification, for better or worse.

Can genetics or anthropology untangle this troubling snarl of conflicting beliefs, incomplete information, and moral quandary? Not yet.

As Wasserman and Walters predicted, new claims for genetic influences on behavioral traits proliferate daily. While there has been a tattered history of such claims, many of which are refuted by further studies, human geneticists are proceeding more cautiously and their results seem less likely to fail.

For example, in June of 1993, a team of geneticists headed by Han Brunner announced the results of a study of several generations of a Dutch family in which the males were prone to extraordinary outbursts of violent or aggressive criminal behavior.[3] Brunner collected information about 15 of the men, all of whom were slightly retarded. Their aberrant behaviors included arson following the deaths of close relatives; attempting to run over an employer who had criticized the aggressor's work; raping a sister; attacking the warden of a mental institution with a pitchfork; threatening a sister at knifepoint. The lack of similar behaviors among the women of the family suggested that the genetic defect, if there were one, was sex-linked. Sex-linked traits are recessive traits encoded on the X chromosome; they are more likely to be expressed in males because men have only one X chromosome. The advantage of working with sex-linked traits like this one is that the gene must reside on the part of the X chromosome that is missing in the Y chromosome.

After ten years of work, Brunner and his colleagues zeroed in on one of two genes that code for an enzyme called monoamine oxidase A (MAO-A) as the most probable mutation. This enzyme breaks down particular chemical signals in the body, especially the neurotransmitters involved in stressful or threatening circumstances. Lending support to the team's hypothesis was their finding that these men excrete unusually high amounts of these neurotransmitters. An American collaborator is trying to clone MAO-A genes from these abnormal patients, as she has already done from normal individuals. The difference between them will pinpoint which gene is involved and exactly what effect the mutation produces.

The link between behavior and gene is easily imagined.[4] Because an MAO-A defect would cause these neurotransmitters to accumulate in abnormal amounts—as they seem to do in the affected men—then any normally stressful circumstance might be perceived as an overwhelmingly threatening situation by individuals with this defect. Their violent responses can then be seen not as unprovoked but as provoked inappropriately, in a classic example of the "hair trigger." Since some foods (chocolate, red wine, some cheeses) carry natural mimics of the neurotransmitters in question, particular diets might produce abnormally high circulating levels of neurotransmitters that might possibly heighten the men's sensitivity to stressful conditions.

If the existence of the postulated MAO-A defect is confirmed by cloning experiments, this study may be one of the first in which a complex behavior is linked firmly to a specific genetic mutation. Both direct evidence and understanding of the mechanism seem close at hand. It is just one family, a single potent example, that provides that link, as is generally the case in genetics. No one has suggested that all aggressive or violent behavior can be attributed to MAO-A defects; no one has hinted that Caucasoids may have a higher-than-average incidence of such defects, simply because this family is (presumably) Caucasoid. And no one has yet explored the strength of cultural, economic, or environmental—especially dietary—influences on the expression of the aggressive behavior or on the social problems caused by it. Is punishment, psychotherapy, dietary regulation, sedation, or medication to supply the needed enzyme the most appropriate response? Wasserman's conference might have proved timely indeed, if only it had been held.

A month after the Dutch team's work was published, another study tackled another apparently sex-linked trait in similar fashion. Dean Hamer and a team from the National Cancer Institute reported strong evidence that male homosexuality is influenced by a gene on a specific, small region of the X chromosome.[5] Their genetic study tends to confirm earlier studies based on other types of research. For example, heritability studies had already indicated a higher-than-expected incidence of homosexuality in males whose twin brothers were gay,[6] and anatomical research had documented differences in the brains of homosexual and heterosexual males.[7]

Hamer and his colleagues were pursuing the genetic link itself, using time-tested techniques. Pedigrees of 76 homosexual men and their families showed more gay male relatives on the maternal than on the paternal side of the family and more homosexual males on the maternal side of the family than in the population at large. Both facts suggested a sex-linked trait.

They then analyzed the DNA of 40 pairs of gay brothers and, for 33 of them, found common genes in a particular region of the X chromosome. Such a result is robust evidence (a technical measure of probability gives a value equivalent to 99.5 percent certainty) that a gene in this region is responsible. The gene itself and its function are as yet unknown. Equally important is the fact that the 7 other pairs of homosexual brothers *do not* seem to possess whatever gene is shared by the 33 pairs, a finding that points to multiple causes for the behaviors lumped together as "homosexual." Indeed, Hamer specifically cautions that the most reasonable interpretation is that male homosexuality has a variety of causes, only one or some of which are genetic.[8]

Once again, the emerging pattern is one of genetic influence over a complex behavior rather than genetic control. Once again, the myriad questions about society's response to an individual's expression of his or her genetic heritage remain unanswered. Since in this instance the gene itself remains elusive, there has been as yet no discussion of its distribution among human populations.

What is the value of the differences? We must decide.

Paleoanthropologists have shown that variation among geographical populations of humans and our ancestors is ancient. If the interpretation of the fossil record were simple, it might shed strong light on the value of the differences among humans. But it is not an easy task to understand the fossil record, which remains fragmentary, sometimes precisely and sometimes poorly dated. Despite remarkable finds in the past few decades, the tens of thousands of fossils are but a tiny sampling of all of the humans and human ancestors that have ever lived.

For these reasons, the history of human variability remains a much-debated issue. Weidenreich and Coon were thwarted by the difficulty of unraveling the origin of races and much criticized for their efforts. Decades later, with a much richer sample of human fossils, the origin of races ought to be much, much clearer, but it is still furiously controversial.

Anthropologists generally agree that the oldest known evidence of our genus and species comes from Africa. As Christopher Stringer, one of the main protagonists in ongoing debates, says, "We are all Africans under the skin."[9] Rather than being the latecomers to sapiensation, as Coon maintained, current evidence suggests that the Negroid peoples of the world may well have been the first to achieve anatomical modernity. Some challenge the African data and suggest an early Asian origin for modern human features. The disagreement is as much about how many modern features make a modern human as it is about unsatisfactory dating of fragmentary fossils.

Alan Thorne and Milford Wolpoff are the foremost proponents of multiregionalism, a reworked version of Weidenreich's and Coon's hypotheses. They are confident that some physical features that distinguish the major, modern racial groups—the Mongoloids, Negroids, Caucasoids, and Australoids—can be traced back almost two million years to forms generally known as *Homo erectus* (the species preceding our own); they deduce a genetic continuity from ancient Java to modern Australoids, from ancient Africa to modern Negroids, from ancient China to modern Mongoloids, and from ancient Europe to modern Caucasoids. Yet Wolpoff and Thorne have put some new spins on their observations, which radically change the political import of their theories from that which clung to Weidenreich's and Coon's efforts.

The multiregionalism hypothesis of Wolpoff and Thorne has a markedly inclusive, rather than exclusive, cast. Humanness is not a possession of one group, even originally. They maintain that regional groups of human ancestors evolved into modern humans—not instantaneously, but simultaneously over a prolonged period of time—in various parts of the world. They use an analogy to explain how this might have occurred. The process is "like several individuals paddling in separate corners of a pool; although they maintain their individuality over time, they influence one another with the spreading ripples they raise (which are the equivalent of genes flowing between populations)."[10]

To Thorne and Wolpoff, the differences that characterize modern races are ancient and deep-rooted. In fact, these scientists would overturn the traditional taxonomy of our ancestors and place all fossil specimens since the time that these racial features first arose—everything other anthropologists would call *Homo erectus*—in our own species,

Homo sapiens.[11] Their argument is a technical one, based on procedures for classifying animals, and their conclusion is the logical (if highly controversial) outcome of the anatomical continuity they see in the fossil record. They believe that some of the differences among modern human races are greater than those between *sapiens* and *erectus*. So if *erectus* is not to be subsumed under a greatly expanded *Homo sapiens*, the only alternative is to remove some modern human populations from *Homo sapiens*. "You don't want to do that," Thorne says pointedly, "do you?"[12]

No one—certainly not Wolpoff and Thorne's most vocal critics—would support such a racist proposition.

The alternative to multiregionalism, known as Out of Africa 2 or the Eve hypothesis, is espoused by Stringer and others who believe modern humans arose in Africa and then spread to the rest of the world, replacing other less advanced populations that already lived there. In this interpretation, humanness developed only once and at only one location. The Out of Africa 2 hypothesis emphasizes that all modern humans share a recent common ancestry that makes racial distinctions recent and biologically trivial. Its proponents find little difficulty in distinguishing *Homo erectus* from *Homo sapiens* and shudder at the thought of lumping skulls with cranial capacities of about 850–900 cc (*Homo erectus*) with modern skulls which average 1250 cc and may be as large as 2000 cc. Brain size is not the only issue, but it is an especially potent one to a species with the audacity to label itself "sapiens," or wise.

Who is correct? It depends how you read the skulls. What Stringer and co-author Clive Gamble call "an excellent fossil sequence [in Africa that] records an evolutionary metamorphosis from ancient specimens . . . through transitional ones . . . to early modern forms"[13] is characterized by Thorne and Wolpoff as "sparse, fragmentary, and for the most part poorly dated . . . [including] materials that do not seem to fit the Eve theory."[14] What Wolpoff and Thorne see as a logically indivisible continuum, Stringer and others have no difficulty separating in two. In short, what is transparent to one is obscure to another. All participants in the debate agree on one thing: more fossils and less emotional discourse are needed. Even in the scientific arena, civility and trust must be maintained.

In addition to the fossil evidence, the proponents of the Out of Africa 2 hypothesis cite controversial work at the interface of genetics and anthropology. Starting in the 1960s, Vincent Sarich and the late Allan Wilson adapted biochemical techniques to measure the genetic distances among living species, with an eye toward clarifying relationships where the fossil record of descent and evolution is unclear. Despite initial methodological problems, the field of molecular evolution has gained substantial reliability and credibility over the past two decades.

By far its most important and most universally accepted conclusion is this: that humans are so closely related to chimpanzees—over 99 percent of our genes are held in common with the chimpanzee[15]—that the differences among the human races are swamped by the tremendous genetic unity among them. The percentage of genes that account for the variation with which we are so obsessed is minuscule; less than 1 percent of our genes can possibly differ from one another. How can the differences be anything but trivial when viewed in this perspective?

More recently Wilson and some of his former students, such as Rebecca Cann, began to apply similar techniques for determining genetic distances to the relationships among the human races.[16] Instead of using molecular differences among various proteins, they have focused on a more precise indicator, the DNA carried in the mitochondrion. Mitochondrial DNA (or mtDNA) is believed to be inherited solely through the maternal line. At fertilization, the mother's egg contains many mitochondria (and lots of mtDNA) while the father's sperm does not contribute its mitochondria to the embryo, presumably leaving them behind in its discarded tail.

MtDNA studies have proved difficult due to the subtlety of genetic variability among living humans, the complexity of defining races due to extensive interbreeding among various populations, and the problem of analyzing large data sets to derive "trees" of most probable descent. One researcher calculated that the 147 samples reported in one of the team's studies would yield a staggering 1.68×10^{294} (or 168 followed by 294 zeroes) possible trees.[17] For these reasons, the conclusions of mtDNA studies have met with a deluge of methodological criticisms and challenges, from both geneticists and paleoanthropologists.[18]

And yet, their answers are intriguing and each new refinement in technique produces little net change in two fundamental points. First, the studies, flawed as they may be, suggest a relatively recent and single-population origin for modern human races, although the exact timing of this origin and the methods for calculting it are hotly debated. Because the original modern humans, as recorded by mtDNA, must have been female, this finding gave birth to the name the "Eve hypothesis," even though Cann and others carefully point out that a *population* with similar genes, not an *individual*, must have given rise to modern humans. Second, their favored interpretation of the data on relationships among races is that modern humans originated among African (Negroid) populations. But it is pragmatically impossible to evaluate the many millions of possible trees their data can generate, nor is it clear just how they should be evaluated. Routinely, geneticists seek "maximum parsimony" trees—ones that postulate the smallest number of convergences, parallelisms, or back-mutations—but there is little hard evidence that this is how evolution actually proceeds. Some critics of the mtDNA work have pointed out that the same data can generate trees with non-African roots that are statistically as plausible or even slightly more parsimonious.[19]

Still, the conclusions reached by mtDNA studies of modern races fit neatly with the Out of Africa 2 hypothesis and seriously undercut the credibility of the multiregionalism theory. But every method, every observation, and every assumption in the entire issue, on all sides, has been subjected to scrutiny and criticism—as they should be when the interpretation can so profoundly influence our vision of ourselves. Was Africa (or any other single locale) the birthplace of modern humanity? Are racial differences ancient, the product of two million years of evolution, or relatively recent? Do these differences matter at all in light of the overwhelming genetic unity of humankind?

The level of debate and disagreement among the geneticists and anthropologists wrestling with the origin of races should not be taken as dismaying evidence of incompetency. It is instead a testimony to the importance of the issues under discussion and the paucity of objective information that is at hand with which to evaluate them. Quiet unanimity—complacent agreement—would be a far more dangerous sign in the face of the handicaps under which any studies of race have been conducted in recent decades. Better to struggle honestly with such difficult topics than to acquiesce into ill-founded certainty too soon.

What is to be done?

A small child explores the world as it grows from infancy to adulthood, discovering the difference between me and not-me, finding out his or her gifts for this activity and handicaps in that. This process is not always pleasant. It can be bitterly disappointing to realize, after weeks or years of training and practice, that you will never be an Olympic athlete or a sculptor of proficiency and vision. Even the humbler realities of life, the commonplace goals and sadnesses—my brother has a better voice than I, my sister gets better math grades—can be hard to accept and embrace. But knowledge and acceptance of self is an integral, indeed an *essential*, part of growing up.

We must look at ourselves hard and be unafraid of what we will discover. Although genetic and biological endowments surely vary, the fulfillment of those potentials is even more variable. The Siamese twin of genetic endowment is the responsibility for the squandering of that potential, a tragedy that happens every day in every life. A child with "tall" genes, if malnourished, will nonetheless be stunted; a child blessed with the eye of an artist will never blossom to delight us all with his or her works if the hand is not trained as well. A potential criminal, if such exists, will never terrorize or hurt if that potential is turned aside by altered circumstances and new empowerment within the bounds of society.

And the responsibility is both collective and individual. We are collectively responsible for the appalling conditions under which so many live, for the opportunities not offered, for the chances not received, for the training never made available. These disasters are our fault, which we must acknowledge and redress. But we are also deeply and individually responsible for our own failures of potential: for the practice skipped, for the basic facts or skills not memorized and reinforced, for the opportunities that were too much hard work to seize, for the moral laziness that makes us lower our standards for ourselves to the point of failure. We all know a gifted student who can't be bothered to do homework; it is a bitter waste. The hope is that we have also all seen the student of modest ability who excels through hard work. It is a lesson we must take to heart.

As a species, we must examine ourselves and accept responsibility for our behavior, as does the growing child, or face the awful prospect of never reaching any sort of maturity.

As a species, it is time to grow up.

Notes

Prologue

1. Charles Darwin, *The Descent of Man and Selection in Relation to Sex*, 1874, 2nd ed., p. 189.

Chapter 1

1. Henrietta Litchfield (ed.), *Emma Darwin: A Century of Family Letters 1792–1896*, 1915, Vol. II, p. 254.
2. Charles Darwin, *The Origin of Species by Means of Natural Selection or the Preservation of Favored Races in the Struggle for Life*, 1859, reprinted in 1958, p. 426.
3. C. Darwin, *Descent*, 1874, p. 707.
4. Francis Darwin (ed.), *Charles Darwin: His Life Told in an Autobiographical Chapter and in a Selected Series of His Published Letters*, 1892, reprinted 1958, pp. 183–184.
5. Phillip Appleman (ed.), *Darwin: A Norton Critical Edition*, 1970, pp. 64–65.
6. C. Darwin, *Origin of Species*, 1859, p. 426.
7. R. H. Davis, *The Medieval Warhorse*, 1989, p. 40.
8. C. Darwin, *Origin of Species*, 1859, p. 57.
9. Ibid., p. 429.
10. Ibid., p. 30.
11. Ibid., p. 449.
12. Voltaire, *Candide*, 1759, Ch. 1, in Emily Morison Beck (ed.), *Familiar Quotations: A Collection of Passages, Phrases and Proverbs Traced to Their Sources in Ancient and Modern Literature*, 1980, p. 343.

13. Details are taken from R. L. Marks, *Three Men of the Beagle*, 1991.

14. John Bowlby, *Charles Darwin: A New Life*, 1990, pp. 270 ff.

15. F. Burkhardt and S. Smith (eds.), *The Complete Correspondence of Charles Darwin* (1985–88), Vol. III, p. 250.

16. Ibid., Vol. III, p. 72.

17. Ibid., Vol. III, pp. 252–256.

18. Ibid., Vol. III, p. 264.

19. John W. Judd, *The Coming of Evolution*, 1935, p. 125.

20. F. Burkhardt and S. Smith, *Complete Correspondence*, Vol. IV, p. 270.

21. Francis Darwin (ed.), *Life and Letters of Charles Darwin*, 1891, Vol. I, p. 386.

22. L. G. Wilson, *Sir Charles Lyell's Scientific Journals on the Species Question*, 1970, pp. xlv–xlvii.

23. Information about Wallace's life and work is taken from Arnold Brackman, *A Delicate Arrangement: The Strange Case of Charles Darwin and Alfred Russel Wallace*, 1980, and Lynn Barber, *The Heyday of Natural History: 1820–1870*, 1980.

24. Francis Darwin and A. C. Seward (eds.), *More Letters of Charles Darwin*, 1903, Vol. I, p. 109.

25. Francis Darwin (ed.), *Autobiography*, 1892, p. 189.

26. Ibid., p. 196.

27. Ibid., pp. 197–198.

28. Ibid., p. 198.

29. Ibid., pp. 198–199.

30. Ibid., p. 199.

31. Ibid., pp. 199–200.

32. Ibid., p. 201.

33. Ibid., pp. 208–209.

34. Quoted in Julian Huxley, "Introduction to the Mentor Edition," 1958, of Darwin, *Origin of Species*, p. xi.

35. F. Darwin (ed.), *Autobiography*, 1892, p. 211.

36. Ibid., p. 231.

Chapter 2

1. Leonard Huxley (ed.), *Life and Letters of Thomas Henry Huxley*, 1900, Vol. I, p. 170.

2. Ibid., Vol. I, p. 107.

3. K. M. Lyell (ed.), *Life and Letters of Sir Charles Lyell, Bart.*, 1881, Vol. II, p. 366.

4. F. Darwin (ed.), *Autobiography*, 1892, p. 222.

5. Thomas Henry Huxley, "On the Reception of the '*Origin of Species*,'"

1891, in F. Darwin (ed.), *Life and Letters*, 1891, Vol. I, pp. 553–558.

6. Gavin de Beer (ed.), *Charles Darwin and Thomas Henry Huxley: Autobiographies*, 1974, p. 101.

7. Thomas H. Huxley, "Agnosticism," *Nineteenth Century*, 1889, February, p. v.

8. Ronald W. Clark, *The Huxleys*, 1968, p. 180.

9. G. de Beer (ed.), *Autobiographies*, 1974, pp. 103–104.

10. L. Huxley (ed.), *Life and Letters*, 1900, Vol. I, p. 17.

11. R. Clark, *The Huxleys*, 1968, p. 13.

12. Ibid., p. 11.

13. L. Huxley (ed.), *Life and Letters*, 1900, Vol. I, p. 25.

14. Julian Huxley (ed.), *T. H. Huxley's Diary of the Voyage of the H.M.S. Rattlesnake*, 1935, p. 47.

15. R. Clark, *The Huxleys*, 1968, p. 20.

16. Ibid., p. 34.

17. L. Huxley (ed.), *Life and Letters*, 1900, Vol. I, p. 442.

18. R. Clark, *The Huxleys*, 1968, p. 44.

19. H. F. Osborn, *Impressions of Great Naturalists*, 1924, pp. 18–21.

20. R. Clark, *The Huxleys*, 1968, p. 45.

21. Richard Owen, *The Life of Richard Owen*, 1894, Vol. I, pp. 23–25.

22. L. Barber, *Heyday*, 1980, p. 176.

23. William Swainson, "A Treatise on Taxidermy," in Lardner's *Cabinet Cyclopaedia of Natural History*, London, 1840, p. 76. Quoted in ibid., p. 175.

24. F. Darwin (ed.), *Autobiography*, 1892, pp. 225–226.

25. Thomas Henry Huxley, "The Origin of Species," *Westminster Review*, 1860, p. 17. Reprinted in Thomas Henry Huxley, *Darwiniana: Essays*, 1898, p. 75. For a clear exposition of their differences, see Mario A. di Gregorio, *T. H. Huxley's Place in Natural Science*, 1984, especially pp. 63–65.

26. Charles Darwin, *The Variation of Animals and Plants under Domestication*, 1868, Vol. I, p. 9.

Chapter 3

1. L. Huxley (ed.), *Life and Letters*, 1900, Vol. I, p. 176.

2. F. Darwin (ed.), *Life and Letters*, 1891, Vol. I, p. 255.

3. Thomas Henry Huxley, "The Darwinian Hypothesis," *The Times*, 1859, Dec. 26. Reprinted in Thomas Henry Huxley, *Darwiniana: Essays*, 1898, pp. 19–20.

4. Ibid., p. 20.

5. L. Huxley (ed.), *Life and Letters*, 1900, Vol. I, p. 177.

6. Morse Peckham, "Introduction to *The Origin of Species* by Charles Darwin, a Variorum Text," 1959. Reprinted in P. Appleman (ed.), *Darwin*, 1970, p. 98.

7. F. Darwin (ed.), *Autobiography,* 1892, p. 245.

8. F. Darwin and A. Seward (eds.), *More Letters*, 1903, Vol. I, p. 185.

9. Richard Owen, "Darwin on the Origin of Species," 1860, *Edinburgh Review*, CXI, pp. 251–275. Reprinted in P. Appleman (ed.), *Darwin*, 1970, pp. 295–296.

10. Adam Sedgwick, "Objections to Mr. Darwin's Theory of the Origin of Species (1860)," 1860, *The Spectator*, XXXIII, March 24, pp. 285–286. Reprinted in P. Appleman (ed.), *Darwin*, 1970, p. 294.

11. William Irvine, *Apes, Angels, and Victorians*, 1972, p. 107.

12. Charles Darwin, unpublished letter to Reginald Darwin, April 8, 1875, Darwin Papers, Cambridge: University Library. Quoted ibid., p. 152.

13. F. Darwin (ed.), *Autobiography*, 1892, p. 250.

14. Duncan Cumming, *The Gentleman Savage: The Life of Mansfield Parkyns 1823–1894*, 1987, p. 153.

15. T. H. Huxley, letter to F. Darwin, 1861, in F. Darwin (ed.), *Autobiography*, 1892, p. 252.

16. W. H. Fremantle, notes supplied to F. Darwin (no date), in F. Darwin (ed.), *Autobiography*, 1892, pp. 251–252.

17. L. Huxley, *Life and Letters*, 1900, Vol. I, pp. 184-185. For the suggestion that Wilberforce's remarks were vulgar, see footnote 2 on pp. 183–184.

18. Ibid., pp. 183–184.

19. W. Irvine, *Apes*, 1972, p. 7.

20. F. Darwin (ed.), *Autobiography*, 1892, p. 253.

21. Cyril Bibby, *Thomas Henry Huxley: Scientist, Humanist, and Educator*, 1959, p. 70.

22. C. Darwin, *Origin of Species*, 1859, p. 488.

23. L. Huxley, *Life and Letters*, 1900, Vol. I, p. 223.

24. Ibid., Vol. I, p. 190.

25. Ibid., Vol. I, pp. 193–194.

26. Thomas Henry Huxley, "Preface" to *Man's Place in Nature and Other Anthropological Essays*, 1900, pp. ix–xi.

27. Thomas Henry Huxley, *Evidence as to Man's Place in Nature*, 1863, reprinted in T. H. Huxley, *Man's Place in Nature*, 1900, pp. 77–78.

28. Ibid., p. 92.

29. Ibid., p. 95.

30. Ibid., p. 111.

31. M. diGregorio, *Huxley's Place*, 1984, p. 153.

32. T. H. Huxley, *Man's Place in Nature*, 1900, pp. 146–147.

33. Ibid., p. 148.

34. Ibid., p. 149.

35. Ibid., pp. 152–153.
36. Ibid., p. 209.
37. See Peter Bowler, *The Eclipse of Darwinism*, 1983, *The Non-Darwinian Revolution: Reinterpreting a Historical Myth*, 1988, and *Theories of Human Evolution: A Century of Debate, 1844-1944*, 1989, for cogent discussions of the delayed acceptance of Darwinian evolution.

Chapter 4

1. F. Darwin (ed.), *Autobiography*, 1892, p. 278.
2. Martin Rudwick, *The Meaning of Fossils*, 1972, p. 218.
3. For more discussion, see P. Bowler, *Non-Darwinian Revolution*, 1988.
4. C. Darwin, *Origin of Species*, 1859, p. 449.
5. For an example, see the discussion of the work of Paul Broca in Erik Trinkaus and Pat Shipman, *The Neandertals: Changing the Image of Mankind*, 1993, pp. 84–85.
6. C. Darwin, *Descent*, 1871, p. 707.
7. H. Litchfield (ed.), *Emma Darwin*, 1915, Vol. II, p. 223.
8. Ernst Haeckel, no reference given, cited in Wilhelm Bölsche, *Haeckel: His Life and Work*, 1906, pp. 38–39.
9. Edward Krumbhaar, "The Centenary of the Cell Doctrine," 1939, *Ann. Med. Hist.* (3S) 1:427–437. Cited in Sherwin B. Nuland, *Doctors: The Biography of Medicine*, 1988, p. 324–325.
10. Ernst Haeckel, *The Story of the Development of the Youth: Letters to His Parents 1852-1856*, 1923, p. 166.
11. Ibid., p. 167.
12. W. Bölsche, *Haeckel*, 1906, p. 70.
13. E. Haeckel, *Story of the Development*, 1923.
14. W. Bölsche, *Haeckel*, 1906, p. 133.
15. C. Darwin, *Origin of Species*, 1859, p. 450.
16. Ernst Haeckel, *Die Radiolarien (Rhizopoda radiara)*, 1862. Quoted in W. Bölsche, *Haeckel*, 1906, p. 141. Italics added.
17. Haeckel, *Die Radiolarien*, 1862, pp. 231–232. Quoted in W. Bölsche, *Haeckel*, 1906, pp. 141–143.
18. W. Bölsche, *Haeckel*, 1906, p. 147.
19. Ibid., p. 149.

Chapter 5

1. Marie Rabl (ed.), *Rudolf Virchow: Letters to His Parents 1839–1864*, 1990, p. 23.
2. Ibid., pp. 28–29.

3. Ibid., p. 61.

4. Byron A. Boyd, *Rudolf Virchow: The Scientist as Citizen*, 1991, p. 40.

5. M. Rabl (ed.), *Letters*, 1990, pp. 110–111.

6. Rudolf Virchow, "Die Kritiker der Cellularpathologie," 1860, *Archiv für pathologische Anatomie und Physiologie und für klinische Medizin*, XVIII:5.

7. K. Sudhoff (ed.), *Rudolf Virchow und die deutschen naturforscherversammlungen*, 1922, pp. 62 and 36.

8. Rudolf Virchow, "Cellular-Pathologie," 1855, *Archiv für pathologische Anatomie und Physiologie und für klinische Medizin*, VIII:26.

9. Ibid., p. 4.

10. Rudolf Virchow quoted in S. Nuland, *Doctors*, 1988, p. 336.

11. The term used for this sort of fossil human still often appears in its archaic spelling, Neanderthal, especially in publications by British scholars, and contributes to confusion among Anglophones as to the correct pronunciation of the name. The formal, scientific name for this specimen and others like it was proposed in 1864 and is *Homo neanderthalensis*. Since such taxonomic names cannot be changed, anthropologists must now deal with the awkward fact that the popular and scientific names for the same specimens differ in spelling.

12. Rudolf Virchow, "Transformismus und Descendenz," 1893, *Berliner klinische Wochenschrift*, XXX:1.

13. Thomas Henry Huxley, "Further Remarks upon the Human Remains from the Neanderthal," 1864, *Natural History Review* 1:435.

14. Daniel Gasman, *The Scientific Origins of National Socialism: Social Darwinism in Ernst Haeckel and the German Monist League*, 1971, pp. 6–7. Gasman's book is a fine analysis of Haeckel's importance and influence on the Nazi movement.

15. Ernst Haeckel, *History of Creation*, 1892, Vol. I, 1–2.

16. D. Gasman, *Scientific Origins*, 1971, p. 35.

17. K. Sudhoff (ed.), *Rudolf Virchow*, 1922, p. 290.

18. Gasman, *Scientific Origins*, 1971, pp. 35–37.

19. Ibid., pp. xxi–xxii.

20. Rudolf Virchow, *The Freedom of Science in the Modern State*, 1878, pp. 55–57. Italics in original.

21. Ibid., pp. 60–62.

22. Ibid., pp. 57–63. Capitals in original.

23. Ibid., pp. 63–64.

24. A. C. Haddon, *History of Anthropology*, 1949, p. 28.

25. Rudolf Virchow, "Anthropology in the Last Twenty Years," 1889, *Annual Report of the Board of Regents of the Smithsonian Institution*, p. 557.

26. Ibid., pp. 563–569.

27. Olor Kloh, "Introduction," 1961, to Ernst Haeckel, *Die Welträthsel*, 1901, pp. vii–viii.

28. D. Gasman, *Scientific Origins*, 1971, p. 15.
29. E. Haeckel, *History of Creation*, 1892, Vol. I, p. 319.
30. Ibid., Vol. II, p. 434.
31. Ernst Haeckel, *Wonders of Life*, 1904, p. 390.
32. C. T. Thiele, no title given, in H. Schmidt (ed.), *Was wir Ernst Häckel verdanke*, 1914, Vol. I, p. 262. Quoted in D. Gasman, *Scientific Origins*, 1971, p. 16.
33. D. Gasman, *Scientific Origins*, 1971, pp. 86 ff.
34. E. Haeckel, *Wonders of Life*, 1904, pp. 21, 118. Haeckel seemed particularly struck by this example and also wrote admiringly of it in an earlier book, *History of Creation*, 1892, pp. 175–176: "A remarkable instance of *artificial* selection in man, on a great scale, is furnished by the ancient Spartans, among whom . . . all newly-born children were subject to a careful examination and selection. All those that were weak, sickly, or affected with any bodily infirmity, were killed. Only the perfectly healthy and strong children were allowed to live, and they alone propagated the race. By this means, the Spartan race was not only continually preserved in excellent strength and vigor, but the perfection of their bodies increased with every generation. No doubt the Spartans owed their rare degree of masculine strength and rough heroic valour (for which they are eminent in ancient history) in great measure to this artificial selection."
35. This interpretation of Haeckel's work is derived from D. Gasman, *Scientific Origins*, 1971.

Chapter 6

1. Francis Galton, *Inquiries into the Human Faculty*, 1883, pp. 24–25.
2. Karl Pearson (ed.), *The Life, Letters and Labours of Francis Galton*, 1914–30, Vol. IIIA, p. 348.
3. Ibid., Vol. II, p. 207.
4. Ibid., Vol. I, p. 203.
5. Daniel J. Kevles, *In the Name of Eugenics: Eugenics and the Uses of Human Heredity*, 1985, p. 10. This book is one of several intelligent syntheses of the history of the eugenics movement and the personalities involved. See also C. P. Blacker, *Eugenics: Galton and After*, 1952; L. A. Farrall, *The Origins and Growth of the English Eugenics Movement: 1865–1925*, 1985; Mark Haller, *Eugenics: Hereditarian Attitudes in American Thought*, 1984; Nancy Stepan, *The Idea of Race in Science: Great Britain 1800–1960*, 1982; and George W. Stocking, Jr., *Race, Culture and Evolution*, 1968.
6. D. Kevles, *In the Name*, 1985, p. 7
7. K. Pearson, *Life, Letters*, 1914–30, Vol. IIIA, p. 124.

8. Francis Galton, *Hereditary Genius: An Inquiry into Its Laws and Consequences*, 1869, pp. 37–38.
9. Ibid., p. 1.
10. Anonymous, *A Eugenics Catechism*, pp. 8–9, quoted in D. Kevles, *In the Name*, 1985, p. 91.
11. D. Kevles, *In The Name*, 1985, pp. 38–39.
12. Ibid., p. 74.
13. F. Darwin and A. Seward, *More Letters*, 1903, Vol. I, p. 317.
14. Ibid., Vol. II, p. 43.
15. This thesis is expounded in the biography by Adrian Desmond and James Moore, *Darwin,* 1991.
16. Leonard Darwin, *The Need for Eugenic Reform*, 1926, p. i.

Chapter 7

1. D. Kevles, *In the Name*, 1985, p. 45.
2. Israel Zangwill, *The Melting Pot*, 1908, Act I. In Emily Morison Beck (ed.), *Familiar Quotations*, 1980, p. 706.
3. Madison Grant, *The Passing of the Great Race; or The Racial Basis of European History*, 1923, p. 17.
4. Ibid., pp. 89–90.
5. Ibid., p. 226.
6. Ibid., p. 263.
7. Charles Davenport, *Heredity in Relation to Eugenics*, 1919, p. 216.
8. Ibid., pp. 216–222.
9. Davenport to Joseph F. Gould, Feb. 17, 1914. Charles B. Davenport papers, Gould file. Quoted in D. Kevles, *In the Name*, 1985, p. 47.
10. Davenport to John C. Merriam, April 4, 1930, Charles B. Davenport Papers, Cold Spring Harbor Series #1. Quoted in D. Kevles, *In the Name*, 1985, p. 48.
11. Davenport to Mrs. Harriman, Feb. 21, 1911, Charles B. Davenport Papers, Mrs. E. H. Harriman file. Quoted in Kevles, *In the Name*, 1985, p. 49.
12. Stephen Jay Gould, *The Mismeasure of Man*, 1981, pp. 168–171. Gould gives a good general review of the history of intelligence testing, pp. 165 ff.
13. D. Kevles, *In the Name*, 1985, p. 109.
14. Buck v. Bell, 274 U.S. 205, 207 (1927), cited in S. Gould, *Mismeasure,* 1981, pp. 335–336.
15. *The New York Times*, 1930, March 11, p. 22. Quoted in D. Kevles, *In the Name*, 1985, p. 113.
16. Thomas Henry Huxley, "The Aryan Question and Prehistoric Man,"

1890, *Nineteenth Century*, reprinted in T. H. Huxley, *Man's Place*, 1900, p. 280.

17. Robert N. Proctor, *Racial Hygiene; Medicine under the Nazis*, 1988, p. 17. Also see this book for an excellent and detailed discussion of evolutionary biology in Nazi Germany.

18. Fritz Brennecke (ed.), *Vom Deutschen Volk und seinem Lebensraum: Handbuch für die Schulungsarbeit in der Hitler Jugend*, 1937, pp. 45–47. Translated into English by H. L. Childs, *The Nazi Primer: Official Handbook for Schooling the Hitler Youth*, 1938.

19. A. R. Mourant, "The Blood Groups of Jews," 1959, *Jewish Journal of Sociology* 1:155–176.

20. D. Gasman, *Scientific Origins*, 1971, pp. 159–160.

21. Personal communication, Matt Cartmill to author.

22. Gerhard Heberer, *Ernst Haeckel und seine wissenschaftliche Bedeutung*, 1934, p. 14. Cited in D. Gasman, *Scientific Origins*, 1971, p. 170.

23. R. Proctor, *Racial Hygiene*, 1988, p. 87.

24. Ibid., pp. 120 ff.

25. Ibid., p. 157.

26. See, for example, Figure 37, ibid., between pp. 182 and 183.

27. Ibid., pp. 183–184.

28. Ibid., pp. 177–222, especially pp. 191–193.

29. Peltret, "Der Arzt als Führer und Ersieher," 1935, *Deutsches Ärzteblatt* 65:565–566. Quoted in Proctor, *Racial Hygiene*, 1988, p. 195.

30. Chester G. Vernier, *American Family Laws*, 1931, Vol. I, section 44, pp. 204–209; also the *1938 Supplement* to *American Family Laws*, 1938, pp. 24–25; summarized in M. F. Ashley Montagu, *Man's Most Dangerous Myth: The Fallacy of Race*, 1942, pp. 187–193.

Chapter 8

1. For cogent discussions of this topic, see Nancy Stepan, *The Idea of Race in Science: Great Britain 1800–1960*, 1982, and George W. Stocking, Jr., *Race, Culture, and Evolution*, 1968.

2. William Z. Ripley, *The Races of Europe*, 1899, p. 108.

3. R. Clark, *Huxleys*, 1968, pp. 114–115. This same work provided much of the information about the life and work of Julian Huxley.

4. Julian Huxley and A. C. Haddon, *We Europeans: A Survey of "Racial" Problems*, 1935. Quoted in R. Clark, *Huxleys*, 1968, p. 280.

5. The interpretation of the new synthesis given here relies heavily upon Ernst Mayr, *The Growth of Biological Thought: Diversity, Evolution, and Inheritance*, 1982, and Ernst Mayr and William Provine (eds.), *The Evolutionary Synthesis*, 1980.

6. D. Kevles, *In the Name*, 1985, p. 213.

7. J. B. S. Haldane, *Daedalus, or Science and the Future*, 1924.

8. D. Kevles, *In the Name*, 1985, p. 191.

9. See E. Mayr, *Growth of Biological*, 1982; E. Mayr and W. Provine, *Evolutionary Synthesis*, 1980; and E. Mayr, *One Long Argument*, 1988.

Chapter 9

1. R. Clark, *Huxleys*, 1968, p. 309.

2. Ibid., pp. 310–311.

3. Ibid.

4. Julian Huxley, *New Bottles for Old Wine*, 1957, p. 12.

5. UNESCO, *The Race Question in Modern Science*, 1956, p. 6.

6. Information about Montagu's life is taken from interviews with the author, November 16–17, 1992.

7. A. Montagu, interview with the author. However, recent international and American editions of *Who's Who* give interestingly different accounts of Montagu's parentage. The international edition lists his parents as "Charles Ehrenberg and Mary," while the American gives them as "Charles Ashley and Mary (Plot)" (p. 2316).

8. A. Montagu, interview with the author.

9. A. Montagu, interview with the author.

10. A. Montagu, interview with the author. Although Montagu gives a 1940 date for this last change of name, M. F. Ashley Montagu is the name that appears on *Man's Most Dangerous Myth*, 1942, and on a paper in a conference volume that was published in 1951.

11. A. Montagu, interview with the author. See also, for example, William E. Fagg, "U.N.E.S.C.O.'s New Statement on Race," 1952, *Man*, No. 3, p. 9.

12. M. F. Ashley Montagu, *Dangerous Myth*, 1942, p. ix.

13. A. Montagu, interview with the author.

14. William E. Fagg, "U.N.E.S.C.O. on Race," 1950, *Man*, No. 220, p. 138.

15. Ibid., pp. 138–139.

16. UNESCO, *Statement on Race*, 1950, printed in *Man*, No. 220, pp. 138–139.

17. Ibid.

18. A. Montagu, interview with the author.

19. William E. Fagg, editorial remark, 1951, *Man*, No. 32, p. 17.

20. Henri Vallois, "U.N.E.S.C.O. on Race," 1951, *Man*, No. 28, pp. 15–16. The original letter reads: "Or l'existence de la race chez l'Homme est un fait biologique incontestable. . . . Poussant à l'extrême la tendance minimisatrice de la race, le manifeste déclare finalement (par. 14):'la race est moins un phénomène biologique qu'un mythe social.' Une telle af-

firmation, qui en arrive à mettre en cause l'existence même de la race chez l'Homme, est en contradiction non seulement avec les faits, mais avec tous les paragraphs antérieurs du manifeste!" Translation by author.

21. W. C. Osman Hill, "U.N.E.S.C.O. on Race," 1951, *Man,* No. 30, pp. 16–17.
22. Ibid.
23. William E. Fagg, "Notes," 1951, *Man*, No. 30, p. 17.
24. Don J. Hager, "Race," 1951, *Man*, No. 93, p. 54.
25. K. L. Little, "U.N.E.S.C.O. on Race," 1951, *Man*, No. 31, p. 17.
26. Ibid.
27. Stanley M. Garn, "Race," 1951, *Man*, No. 200, p. 115.
28. A. Montagu, interview with the author.
29. William E. Fagg, "U.N.E.S.C.O. and Race," 1951, *Man*, No. 101, p. 64.
30. William E. Fagg, "U.N.E.S.C.O.'s New Statement on Race," 1952, *Man*, No. 3, p. 9.
31. Ibid.
32. A. Montagu, interview with the author.
33. UNESCO, *Statement on Race*, 1951, printed in *Man*, 1952, No. 125, p. 91.
34. Ibid.

Chapter 10

1. Earnest A. Hooton, *Apes, Men and Morons*, 1937, p. 152.
2. Ibid., pp. 192 ff.
3. Ibid., p. 210.
4. Ibid., p. 152.
5. A. Montagu, interview with the author.
6. See William Herbert Sheldon and S. S. Stevens, *The Varieties of Temperament: A Psychology of Constitutional Differences,* 1970 reprint of 1942 volume, pp. 1–23, for a thorough description of the project and techniques.
7. Ibid., pp. 13 ff.
8. Ibid., p. 11.
9. Ibid., p. 425.
10. Ibid., p. 428.
11. William H. Sheldon, "The Somatotype, the Morphophenotype and the Morphogenotype," 1951, in Katherine Brehme Warren (ed.), *Origin and Evolution of Man*, Vol. XV, Cold Spring Harbor Symposia on Quantitative Biology, p. 374.
12. A. Montagu, interview with the author; also author's recollection of remarks by Coon, ca. 1977. On this occasion, another paleoanthropologist, Alan Walker, remarked that Montagu had recently visited him in his office. Coon's response was: "You had Ashley Montagu in your

office? And you didn't shoot him?"

13. A. Montagu, interview with the author.

14. Author's recollection of conversation with Coon ca. 1977.

15. A. Montagu, interview with the author.

16. Theodosius Dobzhansky, *Mankind Evolving*, 1962, p. xii.

17. One of the few written sources of information about Dobzhansky's personal life is Theodosius Dobzhansky, *Oral History Memoir*, 1962. See also William Provine, "Origins of the Genetics of Natural Populations Series," 1981, in R. C. Lewontin, John A. Moore, William B. Provine, and Bruce Wallace (eds.), *Dobzhansky's Genetics of Natural Populations, I–XLIII*, pp. 1–83.

18. T. Dobzhansky, *Oral Memoir*, 1962, p. 399.

19. W. Provine, "Origins," 1981, p. 48.

20. A. Montagu, interview with the author.

21. Carleton S. Coon, *Adventures and Discoveries: The Autobiography of Carleton S. Coon, Anthropologist and Explorer*, 1981, p. 204. This book is also the main source for biographical information about Coon.

22. D. Kevles, *In the Name*, 1985, p. 199.

23. M. Demerec, "Foreword," 1951, in K. Warren (ed.), *Origin and Evolution*, p. v.

24. Sherwood L. Washburn, "The Analysis of Primate Evolution with Particular Reference to the Origin of Man," 1951, in K. Warren (ed.), *Origin and Evolution*, p. 67.

25. Ernst Mayr, "Taxonomic Categories in Fossil Hominids," 1951, in K. Warren (ed.), *Origin and Evolution*, p. 114.

26. M. F. Ashley Montagu, "Discussion," 1951, in K. Warren (ed.), *Origin and Evolution*, p. 117.

27. Carleton S. Coon, "Human Races in Relation to Environment and Culture with Special Reference to the Influence of Culture upon Genetic Changes in Human Population," 1951, in K. Warren (ed.), *Origin and Evolution*, p. 249.

28. A. Montagu, in an interview with the author, neither remembered Coon's presence at the meeting nor the content of his paper. This seems to confirm the impression, gleaned from the conference volume, that Coon's paper attracted little attention or comment at the time.

29. M. F. Ashley Montagu, "A Consideration of the Concept of Race," 1951, in K. Warren (ed.), *Origin and Evolution*, pp. 315–336.

30. Carl C. Seltzer, "Constitutional Aspects of Juvenile Delinquency," 1951, in K. Warren (ed.), *Origin and Evolution*, p. 363.

31. Ibid., pp. 366–367.

32. Ibid., p. 370, italics in original.

33. William H. Sheldon, "The Somatotype, the Morphophenotype and the Morphogenotype," 1951, in K. Warren (ed.), *Origin and Evolution*, p. 373.

34. M. F. Ashley Montagu, "Discussion", 1951, in K. Warren (ed.), *Origin and Evolution*, p. 379.
35. W. Sheldon, "Discussion," 1951, in K. Warren (ed.), *Origin and Evolution*, p. 380.
36. Theodosius Dobzhansky, "Human Diversity and Adaptation," 1951, in K. Warren (ed.), *Origin and Evolution*, p. 385. Italics added.

Chapter 11

1. Sherwood Washburn, no reference given, cited in Emöke J. E. Szathmary, "Biological Anthropology," 1992, in Wenner-Gren Foundation for Anthropological Research, *Report for 1990 and 1991: Fiftieth Anniversary Issue*, 1992, p. 21.
2. Ibid., p. 22.
3. Theodosius Dobzhansky, *Mankind Evolving*, 1962, p. 281. Italics in original.
4. Ibid., p. 283.
5. Ibid., pp. 299-300.
6. Ibid., p. xiii.
7. C. Coon, *Adventures*, 1981, p. 356.
8. Lita Osmundsen, interview with author.
9. T. Dobzhansky, *Mankind Evolving*, 1962, pp. 99–100.
10. See, for example, C. Hall and G. Lindzey, *Theories of Personality*, 1957; R. Newman, "Age Changes in Body Build," 1952, *American Journal of Physical Anthropology*, 10:75–90; or J. P. Chaplin and T. S. Krawiec, *Systems and Theories of Psychology*, 1960.
11. See C. Coon, *Adventures*, 1981, pp. 334 ff., for this account and quotations.
12. See S. Gould, *Mismeasure*, 1981, for a discussion.
13. C. Coon, *Adventures*, 1981, p. 335.
14. Ibid.
15. Ibid.
16. Ibid.
17. The relationship between Coon and Putnam was brought up by at least five anthropologists interviewed during the 1990s.

Chapter 12

1. Congoid was Coon's term for the race that included both Negroes and African pygmies. See Carleton S. Coon, *Origin of Races,* 1962, pp. 3–4.
2. Franz Weidenreich, "Facts and Speculations Concerning the Origin of *Homo sapiens*," 1947, *American Anthropologist*, n.s. vol. 49 (2):190.

3. Franz Weidenreich, "The 'Neanderthal Man' and the Ancestors of *Homo sapiens*," 1943, *American Anthropologist*, 45:39.

4. It is sometimes overlooked that Weidenreich specifically mentions the possibility of interbreeding (F. Weidenreich, "Facts and Speculations," 1947, p. 202), and confirms the conspecific status of all modern humans (p. 189). In an earlier paper (F. Weidenreich, " 'Neanderthal Man,' " 1943, p. 39), Weidenreich affirms the unity of ancient humans thus: "If we regard the existing geographical variations of modern mankind as subspecies or races of one main species, the ancestral stages must have had the same taxonomic value." See also Franz Weidenreich, "Interpretations of the Fossil Material," 1949, in W. W. Howells (ed.), *Ideas in Human Evolution; Selected Essays 1949–1961*, 1967, p. 470.

5. C. Coon, *Origin*, 1962, p. 28.

6. Ibid., pp. 29–30.

7. Ibid., p. 481.

8. Ibid., pp. 521-522.

9. The evidence for human origins is reviewed in Fred H. Smith and Frank Spencer (eds.), *Origins of Modern Humans: A World Survey of the Fossil Evidence*, 1984, and P. Mellars and C. B. Stringer (eds.), *The Human Revolution: Behavioral and Biological Perspectives on the Origins of Modern Humans*, 1989.

10. C. Coon, *Origin*, 1962, p. 658.

11. Personal recollection of the author ca. 1975, when a skull known as KNM-ER 1470 was found by Richard Leakey's expedition in Kenya. The skull was believed at the time to be 2.6 million years old, although this proved to be a dating error and the fossil was only about 1.8 million years old. Prior to the publication of cranial capacity of this specimen, Coon had calculated it using a rather Rube Goldberg approximation calibrated by the estimated width of the tie worn by Leakey, who was shown holding the skull in a publicity photo. Coon was disappointed to find that the cranial capacity of this African skull was substantially smaller than his estimate.

12. C. Coon, *Origin*, 1962, p. 663.

13. Ibid., p. 656.

14. For an updated statement of the multiregionalism hypothesis, see Alan G. Thorne and Milford H. Wolpoff, "The Multiregional Evolution of Humans," 1992, *Scientific American* 266 (4):76–83. The attendant controversy is presented in Michael H. Brown, *The Search for Eve*, 1990, and in volume 95 (1) of *American Anthropologist*, published in 1993. In the summer of 1992, the multiregional hypothesis was hotly debated at the Third International Congress on Human Paleontology, and the demise of the major opposing theory, known as the Eve hypothesis, was announced (prematurely) in accounts in the popular press, such as Her-

bert M. Watzman, "Question of Human Origins Debated Anew as Scientists Put to Rest the Idea of a Common African Ancestor," *Chronicle of Higher Education*, Sept. 16, 1992, pp. A7–A8.

15. Ashley Montagu, interview with the author.
16. Theodosius Dobzhansky, "A Debatable Account of the Origin of Races," 1963, *Scientific American* 208 (2):172.
17. Theodosius Dobzhansky, "Genetic Entities in Hominid Evolution," 1963, in Sherwood L. Washburn (ed.), *Classification and Human Evolution*, p. 361.
18. Lita Osmundsen, interview with the author.
19. Theodosius Dobzhansky, "More Bogus 'Science' of Race Prejudice," 1968, *Journal of Heredity* 59 (2):104.
20. Carleton S. Coon, "Comment on 'Bogus Science,' " 1968, *Journal of Heredity* 59 (5):275.
21. Sidney Mintz, interview with the author.
22. Sidney Mintz, letter to Carey Williams, November 30, 1964.
23. F. Clark Howell, interview with the author.
24. C. Coon, *Adventures*, 1981, p. 358.
25. Sherwood L. Washburn, "The Study of Race," 1963, *American Anthropologist*, 65:525.
26. Ibid., pp. 524–527.
27. C. Coon, *Adventures*, 1981, pp. 358–359.
28. Ibid., pp. 353–355.
29. Ashley Montagu, "What Is Remarkable About Varieties of Man Is Likenesses, Not Differences," 1963, *Current Anthropology* 4 (4):361–364, reprinted (and rewritten in part in stronger language) as "On Coon's *The Origin of Races*," in Ashley Montagu (ed.), *The Concept of Race*, 1964, pp. 228–241. Quotes are from the 1964 version, p. 228.
30. Ibid., p. 229.
31. C. Coon, *Origin*, 1962, p. xxxii.
32. A. Montagu, "On Coon's," 1964, pp. 234–236.
33. Carleton S. Coon with E. Hunt, *The Living Races of Man*, 1965, p. ix.
34. C. Coon, *Adventures*, 1981, pp. 367–368.
35. C. Coon, *Adventures*, 1981, p. 368, quoting Edmund Leach, *New York Review of Books*, full reference not given.
36. Alice M. Brues, quoted in C. Coon, *Adventures*, 1981, pp. 370–371.
37. Sol Tax, letter to Leigh Van Valen, March 15, 1965.
38. Leigh Van Valen, "On Discussing Human Races," 1966, *Perspectives in Biology and Medicine* IX (3):378.
39. Ibid., pp. 382–383.
40. Ibid., footnote p. 383.
41. Harold M. Schmeck, Jr., "Carleton S. Coon Is Dead at 76; Pioneer in Social Anthropology," 1981, *The New York Times*, June 6.

42. Michael Bradley, "Book on Race Used Scientific Findings," 1992, letter to *The New York Times*, August 15.

43. Ashley Montagu, letter to *The New York Times*, 1992, August 29, p. 18.

44. Proposed statement to update the UNESCO (1964) document on biological aspects of race, p. 1.

Chapter 13

1. David Wasserman, interview with the author. Most of the personal and professional information given here about Wasserman was taken from this interview and from David Wasserman, "Genetic Factors in Crime: Findings, Uses, and Implications," 1991, grant proposal submitted to the National Center for Human Genome Research, National Institutes of Health.

2. A readable and clear treatment of DNA fingerprinting and its application to legal problems is given in Joseph Wambaugh, *The Blooding*, 1991.

3. Rochelle Cooper Dreyfuss and Dorothy Nelkin,"The Jurisprudence of Genetics," 1992, *Vanderbilt Law Review* 45(2):316.

4. D. Wasserman, "Genetic Factors," 1991, pp. 14–15.

5. Ibid., pp. 17–18.

6. The primary research is described in P. A. Jacobs, M. Brunton, M. M. Melville, R. P. Brittain, and W. F. McClermont, "Aggressive Behavior, Mental Sub-normality, and the XYY Male," 1965, *Nature* 208:1351–1352, and P. A. Jacobs, W. H. Price, S. Richmond, and R. A. W. Ratcliff, "Chromosome Surveys in Penal Institutions and Approved Schools," 1971, *Journal of Medical Genetics* 8:49–58. See also James Q. Wilson and Richard Herrnstein, *Crime and Human Nature*, 1985, pp. 100-102, for a cogent discussion.

7. According to S. Gould, *Mismeasure*, 1981, p. 144, Speck was not an XYY male but a normal XY.

8. See D. Borgaonkar and S. Shah, "The XYY Chromosome, Male—or Syndrome," 1974, *Progress in Medical Genetics* 10:135–222; S. L. Chorover, *From Genesis to Genocide*, 1979; R. Pyeritz, H. Schrier, C. Madansky, L. Miller, and J. Beckwith, "The XYY Male: The Making of a Myth," 1977, in *Biology as a Social Weapon*, pp. 86–100.

9. Genome Study Section, "Summary Statement on Application Number 1 R13 HG00703-01," 1991, pp. 2–3.

10. D. Wasserman, interview with the author.

11. Ibid.

12. Louis Sullivan, quoted in Fox Butterfield, "Dispute Threatens U.S. Plan on Violence," 1992, *The New York Times*, October 23.

13. Louis Sullivan, quoted in "New Federal Center Seeks to Reduce Vio-

lent Crime," 1992, *The New York Times*, June 26, p. A18.

14. Frederick Goodwin, transcript of speech, Feb. 11, 1992, meeting of the National Mental Health Advisory Council, pp. 119–120.

15. See, for example, Barbara Smuts, *Love and Friendship Among Baboons*, 1985.

16. John Conyers, Jr., "A Racially Offensive Attitude," 1992, *Wall Street Journal*, April 1.

Chapter 14

1. Information about Walters from Ronald Walters, interview with the author.

2. R. Walters, interview with the author.

3. Information about Breggin's life and work taken from P. Breggin and G. Ross Breggin, interviews with the author.

4. P. Breggin, quoted in Utrice C. Leid, "Inner City Children Targeted for 'Intervention,'" 1992, *The City Sun* 10(2):4.

5. P. Breggin, interview with the author.

6. Ibid.

7. Paul McHugh, interview with the author.

8. P. Breggin, interview with the author.

9. Letterhead of the Center for the Study of Psychiatry.

10. P. Breggin, interview with the author.

11. P. McHugh, interview with the author.

12. P. Breggin, interview with the author.

13. Ibid.

14. Ibid.

15. Peter Breggin, quoted in U. C. Leid, "Inner-City Children Targeted for 'Intervention,' " 1992, *The City Sun* 10(2):4.

16. Peter Breggin, transcript of testimony before the Violence Initiative Panel, The Congressional Black Caucus Legislative Weekend, 1992, Friday, September 25, p. 3.

17. P. Breggin, interview with the author.

18. Darrel Regier, quoted in Richard Stone, "HHS 'Violence Initiative' Caught in Cross-Fire," 1992, *Science*, 258:213.

19. Louis Sullivan, quoted in Fox Butterfield, "Dispute Threatens U.S. Plan on Violence," 1992, *The New York Times*, October 23; see also "Sullivan Says Research on Violence Isn't Racist," 1992, *The Washington Times*, October 23.

20. Frederick K. Goodwin, "Conduct Disorder as a Precursor to Adult Violence and Substance Abuse: Can the Progression be Halted?" 1992. Transcript of paper delivered before the American Psychiatric Association, Washington, D.C., pp. 2–8.

21. Ibid., p. 9.
22. Karen Schneider, "Study to Quell Violence Is Racist, Critics Charge," 1992, *Detroit Free Press*, November 2, p. 1.
23. *Daily Challenge*, August 18, 1992, cover.
24. Editorial, "Urban Quackery: A Plan for Violent Crime," 1992, *The Arizona Republic*, July 25.
25. *The Final Call*, 1992, Volume 11, Number 24, October 26, cover.
26. D. Wasserman, interview with the author.
27. Ibid.
28. David Wasserman, brochure "Genetics Factors in Crime: Findings, Uses, and Implications," 1992, p. 1.
29. D. Wasserman, interview with the author.
30. D. Wasserman, brochure, p. 2.
31. Ibid., p. 3.
32. P. Breggin, interview with the author.
33. D. Wasserman, interview with the author.
34. P. Breggin, interview with the author.
35. R. Walters, interview with the author.
36. Ronald Walters, "The Politics of the Federal Violence Initiative," 1993, paper delivered to the American Association for the Advancement of Science, February, p. 26.
37. Ibid., p. 25.
38. Ibid., p. 22.
39. Ibid., p. 29.
40. Ibid., p. 26.
41. P. Breggin, letter to editor, *The New York Times*, September 18, 1993, p. A34.
42. R. Walters, interview with the author.
43. This account is taken from David Wasserman, interview with the author; some details were confirmed by Peter Breggin in an interview with the author, but others were disputed. Breggin says he overheard his name in a conversation at Wasserman's table and introduced himself.
44. D. Wasserman, interview with the author.
45. P. Breggin, interview with the author.
46. Alice H. Thomas and Elke Jordan, letter to David Wasserman, July 20, 1992.
47. D. Wasserman, interview with the author.
48. David L. Wheeler, "U. of Md. Conference That Critics Charge Might Foster Racism Loses NIH Support," 1992, *Chronicle of Higher Education*, September 2, p. A6.
49. Ronald Walters, letter to Bernadine Healey, August 18, 1992.
50. John W. Diggs, letter to Victor Medina, August 18, 1992.
51. John W. Diggs, letter to Jacob Goldhaber, September 4, 1992.
52. Bernadine Healey, quoted in "UM to Appeal Loss of Funds for Genet-

ics-Crime Study," 1993, *The Evening Sun*, April 16, p. 1.
53. Eliot Marshall, "NIH Told to Reconsider Crime Meeting," 1993, *Science* 262, p. 23.
54. Ibid, p. 24.
55. R. Walters, interview with the author.
56. Ibid.
57. Ibid.
58. D. Wasserman, interview with the author.
59. S. Gould, *Mismeasure*, 1981, p. 155.
60. Arthur Jensen, "How Much Can We Boost IQ and Scholastic Achievement?" 1969, *Harvard Educational Review* 39 (1):1–123.
61. For a clear and concise review of the significance of heritability estimates, the entire challenge to Jensen's claims, and the difficulties of classifying modern humans into races, see Phillip V. Tobias, *The Meaning of Race*, 1972.

Epilogue

1. See P. V. Tobias, *Meaning of Race*, 1972, pp. 29–35.
2. C. S. Coon with E. Hunt, *The Living Races of Man*, 1965.
3. Han Brunner, M. R. Nelen, P. van Zandvoort, N. G. G. M. Abeling, A. H. van Gennip, E. C. Wolters, M. A. Kulper, H. H. Ropers, and B. A. van Oost, "X-Linked Borderline Mental Retardation with Prominent Behavioral Disturbance: Phenotype, Genetic Localization, and Evidence for Disturbed Monamine Metabolism," 1993, *American Journal of Human Genetics* 52:1032–1039.
4. Virginia Morrell, "Evidence Found for a Possible 'Aggression Gene,' " 1993, *Science* 260 (5115):1722–1723.
5. Dean Hamer, Stella Hu, Victoria L. Magnuson, Nan Hu, and Angela M. L. Pattatucci, "A Linkage Between DNA Markers on the X Chromosome and Male Sexual Orientation," 1993, *Science* 261 (5119):321–327.
6. J. M. Bailey and R. C. Pillard, "Heritable Factors Influence Sexual Orientation in Women," *Archives of General Psychiatry*, 1991, 48:1089.
7. S. LeVay, "Differences in Hypothalamic Structure Between Heterosexual and Homosexual Men," 1991, *Science*, 253:1034; for a popular treatment, see S. LeVay, *The Sexual Brain*, 1993.
8. D. Hamer et al., 1993, "A Linkage," p. 326.
9. Christopher Stringer, interview with the author.
10. Alan G. Thorne and Milford H. Wolpoff, "The Multiregional Evolution of Humans," 1992, *Scientific American*, 266 (4):76.
11. Milford H. Wolpoff, Alan G. Thorne, Jan Jelinek, and Zhang Yinyun, "The Case for Sinking *Homo erectus*: 100 Years of Pithecanthropus Is Enough!" 1991, paper delivered at conference "100 Years of Pithecan-

thropus," Senckenberg, Frankfurt am Main. In press in Jens Franzen (ed.), *100 Years of Pithecanthropus—The* Homo erectus *Problem*.

12. Alan Thorne, interview with the author.

13. Christopher Stringer and Clive Gamble, *In Search of Neanderthals*, 1993. p 136.

14. A. G. Thorne and M. H. Wolpoff, "Multiregional Evolution," 1992, p. 79.

15. Mary-Clare King and Allan Wilson, "Evolution at Two Levels in Humans and Chimpanzees," 1975, *Science*, 188:107–116.

16. The most accessible presentations of the Eve hypothesis are given in A. Wilson and R. Cann, "The Recent African Genesis of Humans," 1992, *Scientific American*, 266 (4):68–73, and Michael H. Brown, *The Search for Eve*, 1990. The most important technical papers are Rebecca Cann, Mark Stoneking, and Allan Wilson, "Mitochondrial DNA and Human Evolution," 1987, *Nature* 325:32–36; R. Cann, "DNA and Human Origins," 1988, *Annual Review of Anthropology* 17:127–143; and L. Vigilant, M. Stoneking, H. Harpending, K. Hawkes, and A. C. Wilson, "African Populations and the Evolution of Human Mitochondrial DNA," 1991, *Science* 253:1503–1507.

17. Alan R. Templeton, "The 'Eve' Hypotheses: A Genetic Critique and Reanalysis," 1993, *American Anthropologist* 95 (1):52.

18. Most of the debate is summarized in four articles: Leslie C. Aiello, "The Fossil Evidence for Modern Human Origins in Africa: A Revised View," 1993, *American Anthropologist*, 95(1):73–96; David W. Frayer, Milford H. Wolpoff, Alan G. Thorne, Fred H. Smith, and Geoffrey Pope, "Theories of Modern Human Origins: The Paleontological Test," 1993, *American Anthropologist,* 95(1):14–50; Robert W. Sussman, "A Current Controversy in Human Evolution," 1993, *American Anthropologist* 95(1):9–14; Alan R. Templeton, "The 'Eve' Hypotheses: A Genetic Critique and Reanalysis," 1993, *American Anthropologist*, 95(1):51–72. Another good source is Christopher Stringer and Clive Gamble, *In Search of Neanderthals*, 1993.

19. S. B. Hedges, S. Kumar, K. Tamura, and M. Stoneking, "Human Origins and Analysis of Mitochondrial DNA Sequences," 1992, *Science* 255:737–739; D. R. Maddison, M. Ruvulo, and D. L. Swofford, "Geographic Origins of Human Mitochondrial DNA: Phylogenetic Evidence from Control Region Sequences," 1992, *Systematic Biology,* 41:111–124.

Bibliography

MANUSCRIPTS AND PERSONAL PAPERS

American Association of Physical Anthropologists. "Proposed Statement
    to Update the UNESCO (1964) Document on Biological Aspects of
    Race," 1993, pp. 1–2.
Breggin, Peter. Transcript of testimony before the Violence Initiative Pan-
    el, The Congressional Black Caucus Legislative Weekend, September
    25, 1992.
Diggs, John W. Letter to Victor Medina, August 18, 1992.
———. Letter to Jacob Goldhaber, September 4, 1992.
Genome Study Section, National Institutes of Health. "Summary Statement
    on Application Number 1 R13 HG00703-01," 1991.
Goodwin, Frederick K. "Conduct Disorder as a Precursor to Adult Violence
    and Substance Abuse: Can the Progression Be Halted?" Transcript of
    paper delivered before the American Psychiatric Association, Wash-
    ington, D.C., 1992.
———. Transcript of speech, Feb. 11 meeting of the National Mental
    Health Advisory Council, 1992.
Mintz, Sidney. Letter to Carey Williams, November 30, 1964.
Tax, Sol. Letter to Leigh Van Valen, March 15, 1965.
Thomas, Alice H., and Elke Jordan. Letter to David Wasserman, July 20,
    1992.
Walters, Ronald. Letter to Bernadine Healey, August 18, 1992.
———. "The Politics of the Federal Violence Initiative." Paper delivered
    to the American Association for the Advancement of Science, Febru-
    ary, 1993.
Wasserman, David. "Genetic Factors in Crime: Findings, Uses, and Impli-
    cations." Grant proposal submitted to the National Center for Human

Genome Research, National Institutes of Health, 1991.
———. "Genetic Factors in Crime: Findings, Uses, and Implications," brochure, 1992.

BOOKS, ARTICLES, AND DISSERTATIONS

Ackerknecht, E. *Rudolf Virchow: Doctor, Statesman, Anthropologist.* Madison: University of Wisconsin Press, 1953.

Aiello, Leslie. "The Fossil Evidence for Modern Human Origins in Africa: A Revised View." *American Anthropologist* 95(1):73–96, 1993.

Anonymous. *A Eugenics Catechism.* American Eugenics Society, 1926.

———. "New Federal Center Seeks to Reduce Violent Crime." *The New York Times*, June 26, 1992, p. A18.

———. *Daily Challenge.* August 18, 1992, cover.

———. "Natural Criminals?" *The Final Call* 11(24): cover, 1992.

———. "Sullivan Says Research on Violence Isn't Racist." *The Washington Times*, October 23, 1992.

———. "Urban Quackery: A Plan for Violent Crime." *The Arizona Republic*, July 25, 1992.

———. "UM to Appeal Loss of Funds for Genetics-Crime Study." *The Evening Sun*, April 16, 1993, Baltimore, p. 1.

Appleman, Phillip, ed. *Darwin: A Norton Critical Edition.* New York: W. W. Norton, 1970.

Bailey, J. M., and R. C. Pillard. "Heritable Factors Influence Sexual Orientation in Women." *Archives of General Psychiatry* 48:1089, 1991.

Barber, Lynn. *The Heyday of Natural History: 1820–1870.* Garden City: Doubleday, 1980.

Beck, Emily Morison, ed., and John Bartlett, comp. *Familiar Quotations: A Collection of Passages, Phrases and Proverbs Traced to Their Sources in Ancient and Modern Literature*, 15th ed. Boston: Little, Brown, 1980.

Bibby, Cyril. *Thomas Henry Huxley: Scientist, Humanist, and Educator.* London: Watts, 1959.

Blacker, C. P. *Eugenics: Galton and After.* London: Gerald Duckworth & Co., 1952.

Bölsche, Wilhelm. *Haeckel: His Life and Work.* Tr. Joseph McCabe. London: T. Fisher Unwin, 1906.

Borgaonkar, D., and S. Shah. "The XYY Chromosome, Male—or Syndrome." *Progress in Medical Genetics* 10:135–222, 1974.

Bowlby, John. *Charles Darwin: A New Life.* New York: W. W. Norton, 1990.

Bowler, Peter. *The Eclipse of Darwinism.* Baltimore: Johns Hopkins University Press, 1983.

———. *The Non-Darwinian Revolution: Reinterpreting a Historical Myth.* Baltimore: Johns Hopkins University Press, 1988.

———. *Theories of Human Evolution: A Century of Debate, 1844-1944.* Baltimore: Johns Hopkins University Press, 1989.

Boyd, Byron A. *Rudolf Virchow: The Scientist as Citizen.* New York: Garland Publishing Co., 1991.

Brackman, Arnold. *A Delicate Arrangement: The Strange Case of Charles Darwin and Alfred Russel Wallace.* New York: Times Books, 1980.

Bradley, Michael. "Book on Race Used Scientific Findings." *The New York Times*, August 15, 1992.

Breggin, Peter. Letter to Editor. *The New York Times*, September 18, 1993, p. A34.

Brennecke, Fritz, ed. *Vom Deutschen Volk und seinem Lebensrau:, Handbuch für die Schulungsarbeit in der Hitler Jugend.* Munich. 1937. Translated into English by H. L. Childs as *The Nazi Primer: Official Handbook for Schooling the Hilter Youth.* New York, 1938.

Brown, Michael H. *The Search for Eve.* New York: Harper & Row, 1990.

Brunner, Han, M. R. Nelen, P. van Zandvoort, N. G. G. M. Abeling, A. H. van Gennip, E. C. Wolters, M. A. Kulper, H. H. Ropers, and B. A. van Oost. "X-Linked Borderline Mental Retardation with Prominent Behavioral Disturbance: Phenotype, Genetic Localization, and Evidence for Disturbed Monamine Metabolism." *American Journal of Human Genetics* 52:1032–1039, 1993.

Burkhardt, F., and S. Smith, eds. *The Complete Correspondence of Charles Darwin.* Cambridge: Cambridge University Press, 1985–88.

Butterfield, Fox. "Dispute Threatens U.S. Plan on Violence." *The New York Times*, October 23, 1992.

Cann, Rebecca L. "DNA and Human Origins." *Annual Review of Anthropology* 17:127–143, 1988.

Cann, R. L., M. Stoneking, and A. C. Wilson. "Mitochondrial DNA and Human Evolution." *Nature* 325:32–36, 1987.

Chaplin, J. P., and T. S. Krawiec. *Systems and Theories of Psychology.* New York: Holt, Rinehart & Winston, 1960.

Chorover, S. L. *From Genesis to Genocide.* Cambridge: Massachusetts Institute of Technology Press, 1979.

Clark, Ronald W. *The Huxleys.* New York and Toronto: McGraw-Hill, 1968.

Conyers, John, Jr. "A Racially Offensive Attitude." *Wall Street Journal*, April 1, 1992.

Coon, Carleton S. *Adventures and Discoveries: The Autobiography of Carleton S. Coon, Anthropologist and Explorer.* Englewood Cliffs: Prentice-Hall, 1981.

———. "Comment on 'Bogus Science.'" *Journal of Heredity* 59(5):275, 1968.

————. "Human Races in Relation to Environment and Culture with Special Reference to the Influence of Culture upon Genetic Changes in Human Population." In Katherine Brehme Warren, ed., *Origin and Evolution of Man*. Cold Spring Harbor: Cold Spring Harbor Symposia on Quantitative Biology, 1951, Vol. XV: 247–257.

————. *The Origin of Races*. New York: Alfred A. Knopf, 1962.

———— with E. Hunt. *The Living Races of Man*. New York: Alfred A. Knopf, 1965.

Cowan, Ruth Schwartz. *Sir Francis Galton and the Study of Heredity in the Nineteenth Century*. New York: Garland, 1985.

Cumming, Duncan. *The Gentleman Savage: The Life of Mansfield Parkyns 1823–1894*. London: Century, 1987.

Darwin, Charles. *The Descent of Man and Selection in Relation to Sex*, 2nd ed., New York: A. L. Burt Company, 1874.

————. *The Origin of Species by Means of Natural Selection or the Preservation of Favored Races in the Struggle for Life*. London: John Murray & Sons, 1859. Reprinted in 1958, New York: The New American Library.

————. *The Variation of Animals and Plants under Domestication*. London: John Murray & Sons, 1868.

Darwin, Francis, ed. *Charles Darwin: His Life Told in an Autobiographical Chapter and in a Selected Series of His Published Letters*. New York: D. Appleton and Company, 1892. Reprinted as Darwin, Francis, ed., *The Autobiography of Charles Darwin and Selected Letters*, New York: Dover Publications, 1958.

Darwin, Francis, ed. *Life and Letters of Charles Darwin*. London: John Murray & Sons, 1891.

Darwin, Francis, and A. C. Seward, eds., *More Letters of Charles Darwin*. London: John Murray & Sons, 1903.

Darwin, Leonard. *The Need for Eugenic Reform*. New York: D. Appleton, 1926.

Davenport, Charles. *Heredity in Relation to Eugenics*. New York: Henry Holt, 1919.

Davis, R. H. C. *The Medieval Warhorse*. London: Thames and Hudson, 1989.

de Beer, Gavin, ed. *Charles Darwin and Thomas Henry Huxley: Autobiographies*. London: Oxford University Press, 1974.

Demerec, M. "Foreword." In Katherine Brehme Warren, ed., *Origin and Evolution of Man*. Cold Spring Harbor: Cold Spring Harbor Symposia on Quantitative Biology, 1951, Vol. XV:v–vi.

Desmond, Adrian, and James Moore. *Darwin*. New York: Warner Books, 1991.

di Gregorio, Mario A. *T. H. Huxley's Place in Natural Science*. New Haven: Yale University Press, 1984.

Dobzhansky, Theodosius. "A Debatable Account of the Origin of Races." *Scientific American* 208(2):169–172, 1963.

———. "Genetic Entities in Hominid Evolution." In Sherwood L. Washburn, ed., *Classification and Human Evolution*. London: Metheun and Co., 1963, pp. 347–362.

———. "Human Diversity and Adaptation." In Katherine Brehme Warren, ed., *Origin and Evolution of Man*. Cold Spring Harbor: Cold Spring Harbor Symposia on Quantitative Biology, 1951, Vol. XV:385–400.

———. *Mankind Evolving*. New Haven: Yale University Press, 1962.

———. "More Bogus 'Science' of Race Prejudice." *Journal of Heredity* 59(2):102–104, 1968.

———. *Oral History Memoir*. New York: Columbia University Press, 1962.

Dreyfuss, Rochelle Cooper, and Dorothy Nelkin. "The Jurisprudence of Genetics." *Vanderbilt Law Review* 45(2):313–348, 1992.

Fagg, William E. Editorial remark. *Man*, No. 32:17, 1951.

———. "Notes." *Man*, No. 30:17, 1951.

———. "U.N.E.S.C.O. on Race." *Man*, No. 220:138, 1950.

———. "U.N.E.S.C.O. and Race." *Man*, No. 101:64, 1951.

———. "U.N.E.S.C.O.'s New Statement on Race." *Man*, No. 3:9, 1952.

Farrall, L. A. *The Origins and Growth of the English Eugenics Movement: 1865–1925*. New York: Garland, 1985.

Frayer, David, Milford Wolpoff, Alan Thorne, Fred Smith, and Geoffrey Pope. "Theories of Modern Human Origins: The Paleontological Test." *American Anthropologist* 95(1):14–50, 1993.

Galton, Francis. *Hereditary Genius: An Inquiry into Its Laws and Consequences*. London: Macmillan, 1869.

———. *Inquiries into the Human Faculty*. London: Macmillan, 1883.

Garn, Stanley M. "Race." *Man*, No. 200:115, 1951.

Gasman, Daniel. *The Scientific Origins of National Socialism: Social Darwinism in Ernst Haeckel and the German Monist League*. London: MacDonald, 1971.

Gould, Stephen Jay. *The Mismeasure of Man*. New York: W. W. Norton, 1981.

Grant, Madison. *The Passing of the Great Race; or The Racial Basis of European History*, 4th ed. New York: Charles Scribners' Sons, 1923.

Haddon, A. C. *History of Anthropology*. London: G. P. Putnam's Sons, 1949.

Haeckel, Ernst. *Die Radiolarien (Rhizopoda radiara)*, 1862.

Haeckel, Ernst. *The Evolution of Man*. Translation by Joseph McCabe of *Anthropogenie*, 5th ed. New York: D. Appleton, 1905.

———. *The History of Creation*. Translation by E. Ray Lankester of *Natürliche Schöpfungsgeschichte*, 4th ed. London: Kegan Paul, Trench, Trubner & Co., 1892.

———. *The Riddle of the Universe.* Translation by Joseph McCabe of *Die Welträthsel*, 6th ed. New York: D. Appleton, 1901.

———. *The Story of the Development of the Youth: Letters to His Parents 1852–1856.* Translation by G. Barry Gifford. New York: Harper & Brothers, 1923.

———. *The Wonders of Life.* Translation by Joseph McCabe of *Die Lebenswunder*. New York: Harper, 1904.

Hager, Don J. "Race." *Man*, No. 93:54, 1951.

Haldane, J. B. S. *Daedalus, or Science and the Future.* London: E. P. Dutton, 1924.

Hall, C., and G. Lindzey. *Theories of Personality.* New York: Wiley, 1957.

Haller, Mark. *Eugenics: Hereditarian Attitudes in American Thought.* New Brunswick: Rutgers University Press, 1984.

Hamer, Dean, Stella Hu, Victoria L. Magnuson, Nan Hu, and Angela M. L. Pattattucci. "A Linkage Between DNA Markers on the X Chromosome and Male Sexual Orientation." *Science* 261(5119):321–327, 1993.

Heberer, Gerhard. *Ernst Haeckel und seine wissenschaftliche Bedeutung.* Tubingen: Heine, 1934.

Hedges, S. B., S. Kumar, K. Tamura, and M. Stoneking. "Human Origins and Analysis of Mitochondrial DNA Sequences." *Science* 255:737–739, 1992.

Hooton, Earnest A. *Apes, Men and Morons.* New York: Charles Scribners' Sons, 1937.

Huxley, Julian. "Introduction to the Mentor Edition of *Origin of Species.*" New York: New American Library, 1958.

———. *New Bottles for Old Wine.* New York: Harper & Bros., 1957.

———, ed. *T. H. Huxley's Diary of the Voyage of the H.M.S. Rattlesnake.* London: Chatto and Windus, 1935.

Huxley, Julian, and A. C. Haddon. *We Europeans: A Survey of "Racial" Problems.* New York: Harper and Bros., 1936.

Huxley, Leonard, ed. *Life and Letters of Thomas Henry Huxley.* London: Macmillan and Co., 1900.

Huxley, Thomas Henry. "Agnosticism." *Nineteenth Century* 25:169–194, 1889.

———. "The Aryan Question and Prehistoric Man." *Nineteenth Century* 28:750–777, 1890. Reprinted in T. Huxley,. *Man's Place in Nature and Other Anthropological Essays.* New York: D. Appleton, 1900, pp. 272–329.

———. "The Darwinian Hypothesis." *The Times*, Dec. 26, 1859. Reprinted in Thomas Henry Huxley. *Darwiniana: Essays.* New York: D. Appleton and Company, 1898, 1–21.

———. "Further Remarks upon the Human Remains from the Neanderthal." *Natural History Review* 1:429–446, 1864.

———. "The Origin of Species." *Westminster Review* 17:541–570, 1860.

Reprinted in Thomas Henry Huxley, *Darwiniana: Essays*. New York: D. Appleton and Company, 1898, pp. 22–79.

———. "Preface" to *Man's Place in Nature and Other Anthropological Essays*. New York: D. Appleton, 1900, pp. ix–xi.

———. "On the Reception of the 'Origin of Species.' " In F. Darwin, ed., *Life and Letters of Charles Darwin*. London: John Murray & Sons, 1891.

Irvine, William. *Apes, Angels, and Victorians*. New York: McGraw-Hill, 1972.

Jacobs, P. A., M. Brunton, M. M. Melville, R. P. Brittain, and W. F. McClermont. "Aggressive Behavior, Mental Sub-normality, and the XYY Male." *Nature* 208:1351–1352, 1965.

Jacobs, P. A., W. H. Price, S. Richmond, and R. A. W. Ratcliff. "Chromosome Surveys in Penal Institutions and Approved Schools." *Journal of Medical Genetics* 8:49–58, 1971.

Jensen, A. R. "How Much Can We Boost IQ and Scholastic Achievement?" *Harvard Educational Review* 39(1):1–123, 1969.

Judd, John W. *The Coming of Evolution*. Cambridge: Cambridge University Press, 1935.

Kevles, Daniel J. *In the Name of Eugenics: Eugenics and the Uses of Human Heredity*. New York: Alfred A. Knopf, 1985.

King, Mary-Clare, and Allan C. Wilson. "Evolution at Two Levels in Humans and Chimpanzees." *Science* 188:107–116, 1975.

Kloh, Olor. "Introduction" to reprint of Ernst Haeckel. *Die Welträthsel*. Berlin: Akademic Verlag, 1901. Reprinted 1961, pp. vii–viii.

Krumbhaar, Edward. "The Centenary of the Cell Doctrine." *Ann. Med. Hist.* (3S) 1:427–437, 1939.

Leid, Utrice C. 'Inner-City Children Targeted for 'Intervention.' " *The City Sun*, June 17–23, 1992, pp. 4–8.

LeVay, Simon. *The Sexual Brain*. Cambridge: MIT Press, 1993.

———. "Difference in Hypothalamic Structure Between Heterosexual and Homosexual Men." *Science* 253:1034–1035, 1991.

Litchfield, Henrietta, ed. *Emma Darwin: A Century of Family Letters 1972–1896*. London: John Murray & Sons, 1915.

Little, K. L. "U.N.E.S.C.O. on Race." *Man*, No. 31:17, 1951.

Lyell, K. M., ed. *Life and Letters of Sir Charles Lyell, Bart*. London: John Murray & Sons, 1881.

Maddison, D. R., M. Ruvulo, and D. L. Swofford. "Geographic Origins of Human Mitochondrial DNA: Phylogenetic Evidence from Control Region Sequences." *Systematic Biology* 41:111–124, 1992.

Marks, R. L. *Three Men of the Beagle*. New York: Alfred A. Knopf, 1991.

Marshall, Eliot. "NIH Told to Reconsider Crime Meeting." *Science* 262:23–24, 1993.

Mayr, Ernst. *The Growth of Biological Thought: Diversity, Evolution, and*

Inheritance. Cambridge: Belknap Press, 1982.

———. *One Long Argument*. Cambridge: Harvard University Press, 1988.

———. "Taxonomic Categories in Fossil Hominids." In Katherine Brehme Warren, ed., *Origin and Evolution of Man*. Cold Spring Harbor: Cold Spring Harbor Symposia on Quantitative Biology, 1951, Vol. XV:109–117.

Mayr, Ernst, and William Provine, eds. *The Evolutionary Synthesis*. Cambridge: Harvard University Press, 1980.

Mellars, P., and C. B. Stringer, eds. *The Human Revolution: Behavioral and Biological Perspectives on the Origins of Modern Humans*. Edinburgh: University of Edinburgh Press, 1989.

Montagu, Ashley. Letter. *The New York Times*, August 29, 1992, p. 18.

———. "What Is Remarkable About Varieties of Man Is Likenesses, Not Differences." *Current Anthropology* 4(4):361–364, 1963. Reprinted (and rewritten in part in stronger language) as "On Coon's *The Origin of Races*," in Ashley Montagu, ed., *The Concept of Race*. New York: The Free Press, 1964, pp. 228–241.

Montagu, M. F. Ashley. "A Consideration of the Concept of Race." In Katherine Brehme Warren, ed. *Origin and Evolution of Man*. Cold Spring Harbor: Cold Spring Harbor Symposia on Quantitative Biology, 1951, Vol. XV:315–333.

———. "Discussion". In Katherine Brehme Warren, ed., *Origin and Evolution of Man*. Cold Spring Harbor: Cold Spring Harbor Symposia on Quantitative Biology, 1951, Vol. XV:117–118

———. "Discussion." In Katherine Brehme Warren, ed., *Origin and Evolution of Man*. Cold Spring Harbor: Cold Spring Harbor Symposia on Quantitative Biology, 1951, Vol. XV:379.

———. *Man's Most Dangerous Myth: The Fallacy of Race*. New York: Columbia University Press, 1942.

Morrell, Virginia. "Evidence Found for a Possible 'Aggression Gene.' " *Science* 260(5115):1722–1723, 1993.

Mourant, A. R. "The Blood Groups of Jews." *Jewish Journal of Sociology* 1:155–176, 1959.

Newman, R. "Age Changes in Body Build." *American Journal of Physical Anthropology* 10:75–90, 1952.

Nuland, Sherwin B. *Doctors: The Biography of Medicine*. New York: Vintage Books, 1988.

Osborn, Henry Fairfield. *Impressions of Great Naturalists*. New York: Charles Scribners' Sons, 1924.

Osman Hill, W. C. "U.N.E.S.C.O. on Race." *Man*, No. 30:16–17, 1951.

Owen, Richard. "Darwin and the Origin of Species." *Edinburgh Review* CXI, 251–275, 1860. Reprinted in P. Appleman, ed. *Darwin: A Norton Critical Edition*. New York: W. W. Norton, 1970, pp. 295–296.

———. *The Life of Richard Owen*. London: John Murray & Sons, 1894.

Pearson, Karl, ed., *The Life, Letters and Labours of Francis Galton*. Cambridge: Cambridge University Press, 1914–30.

Peckham, Morse. "Introduction to *The Origin of Species* by Charles Darwin, a Variorum Text." Philadelphia: University of Pennsylvania Press, 1959. Reprinted in Philip Appleman, ed., *Darwin: A Norton Critical Edition*. New York: W. W. Norton, 1970, pp. 98–100.

Peltret. "Der Arzt als Führer und Ersieher." *Deutsches Ärzteblatt* 65:565–566, 1935.

Proctor, Robert N. *Racial Hygiene: Medicine under the Nazis*. Cambridge: Harvard University Press, 1988.

Provine, William. "Origins of the Genetics of Natural Populations Series." In R. C. Lewontin, John A. Moore, William B. Provine, and Bruce Wallace, eds., *Dobzhansky's Genetics of Natural Populations*, I–XLIII. New York: Columbia University Press, 1981, pp. 1–83.

Pyeritz, R., H. Schrier, C. Madansky, L. Miller, and J. Beckwith. "The XYY Male: The Making of a Myth." In The Ann Arbor Science for the People Editorial Collective, ed., *Biology as a Social Weapon*. Minneapolis-St. Paul: Burgess Publishing Co., 1977, pp. 86–100.

Rabl, Marie, ed. *Rudolf Virchow: Letters to His Parents 1839–1864*. Translated, revised, and edited by L. J. Rather. Canton, MA: Science History Publications, 1990.

Ripley, William Z. *The Races of Europe*. New York: D. Appleton, 1899.

Rudwick, Martin. *The Meaning of Fossils: Episodes in the History of Palaeontology*. New York: Science History Publications, a division of Neale Watson Academic Publications, 1972.

Schmeck, Harold M., Jr. "Carleton S. Coon Is Dead at 76; Pioneer in Social Anthropology." *The New York Times*, June 6, 1981.

Schneider, Karen. "Study to Quell Violence Is Racist, Critics Charge." *Detroit Free Press*, November 2, 1992, p. 1.

Sedgwick, Adam. "Objections to Mr. Darwin's Theory of the Origin of Species (1860)." *The Spectator* XXXII:85–286, 1860. Reprinted in Philip Appleman, ed., *Darwin: A Norton Critical Edition*. New York: W. W. Norton, 1970, pp. 292–294.

Seltzer, Carl C. "Constitutional Aspects of Juvenile Delinquency." In Katherine Brehme Warren, ed., *Origin and Evolution of Man*. Cold Spring Harbor: Cold Spring Harbor Symposia on Quantitative Biology, 1951, Vol. XV:361–371.

Sheldon, William Herbert. "Discussion." In Katherine Brehme Warren, ed., *Origin and Evolution of Man*, Cold Spring Harbor: Cold Spring Harbor Symposia on Quantitative Biology, 1951, Vol. XV:380.

———. "The Somatotype, the Morphophenotype, and the Morphogenotype." In Katherine Brehme Warren, ed., *Origin and Evolution of Man*. Cold Spring Harbor: Cold Spring Harbor Symposia on Quantitative Biology, 1951, Vol. XV:373–376.

Sheldon, William Herbert, and S. S. Stevens. *The Varieties of Tempera-ment: A Psychology of Constitutional Differences*. New York: Hafner Publishing Company, 1942, reprinted in 1970.

Smith, Fred H., and Frank Spencer, eds. *Origins of Modern Humans: A World Survey of the Fossil Evidence*. Philadelphia: Alan R. Liss, 1984.

Smuts, Barbara. *Love and Friendship Among Baboons*. New York: Aldine Publishing Company, 1985.

Stepan, Nancy. *The Idea of Race in Science: Great Britain 1800-1960*. Hamden, Conn.: Archon Books, 1982.

Stocking, George W., Jr. *Race, Culture, and Evolution: Essays in the His-tory of Anthropology*. New York: Macmillan, 1968.

Stone, Richard. "HHS 'Violence Initiative' Caught in Cross-Fire." *Science* 258:212–213, 1992.

Stringer, Christopher, and Clive Gamble. *In Search of Neanderthals*. Lon-don: Thames and Hudson, 1993.

Sudhoff, K., ed. *Rudolf Virchow und die deutschen Naturforscherver-sammlungen*. Leipzig, 1922.

Sussman, Robert. "A Current Controversy in Human Evolution." *American Anthropologist* 95(1):9–13, 1993.

Swainson, William. "A Treatise on Taxidermy." In Lardner's *Cabinet Cy-clopaedia of Natural History*. London, 1840.

Szathmary, Emöke J. E. "Biological Anthropology." In Wenner-Gren Foun-dation for Anthropological Research. *Report for 1990 and 1991: Fifti-eth Anniversary Issue*. New York, 1992, pp. 18–30.

Templeton, Alan R. "The 'Eve' Hypotheses: A Genetic Critique and Re-analysis." *American Anthropologist* 95(1):57–72, 1993.

Thiele, C. T. In H. Schmidt, ed., *Was wir Ernst Häckel verdanke*. Leipzig: Unesma, 1914.

Thompson, D. G. "Herbert Spencer." In Brooklyn Ethical Association, *Evo-lution: Popular Lectures and Discussions Before the Brooklyn Ethical Association*. Boston: James H. West, 1889, pp. 3–22.

Thorne, Alan. G., and Milford H. Wolpoff. "The Multiregional Evolution of Humans." *Scientific American* 266(4):76–83, 1992.

Tobias, P. V. *The Meaning of Race*, 2nd ed. Johannesburg: South African Institute of Race Relations, 1972.

Trinkaus, Erik, and Pat Shipman. *The Neandertals: Changing the Image of Mankind*. New York: Alfred A. Knopf, 1993.

UNESCO. *The Race Question in Modern Science*. New York: Whiteside Inc. and William Morrow and Co., 1956.

———. *Statement on Race*. 1950. Reprinted in *Man*, No. 220:138–139, 1950.

———. *Statement on Race*. 1951. Reprinted in *Man*, No. 125:91, 1952.

Vallois, Henri. "U.N.E.S.C.O. on Race." *Man*, No. 28:15–16, 1951.

Vernier, Chester G. *American Family Laws*. Palo Alto: Stanford University Press, 1931.

Van Valen, Leigh. "On Discussing Human Races." *Perspectives in Biology and Medicine* IX(3):377–383, 1966.

Vigilant, Linda, M. Stoneking, H. Harpending, K. Hawkes, and A. C. Wilson. "African Populations and the Evolution of Human Mitochondrial DNA." *Science* 253:1503–1507, 1991.

Virchow, Rudolf. "Anthropology in the Last Twenty Years." *Annual Report of the Board of Regents of the Smithsonian Institution*, 1889.

———. "Cellular-Pathologie." *Archiv für pathologische Anatomie und Physiologie und für klinische Medizin* VIII, 1855.

———. "Die Kritiker der Cellularpathologie." *Archiv für pathologische Anatomie und Physiologie und für klinische Medizin* XVIII: 5, 1860.

———. *The Freedom of Science in the Modern State*. London: John Murray & Sons, 1878.

———. "Transformismus und Descendenz." *Berliner klinische Wochenschrift* XXX, 1893.

Wambaugh, Joseph. *The Blooding*. New York: Bantam, 1991.

Washburn, Sherwood L. "The Analysis of Primate Evolution with Particular Reference to the Origin of Man." In Katherine Brehme Warren, ed., *Origin and Evolution of Man*. Cold Spring Harbor: Cold Spring Harbor Symposia on Quantitative Biology, 1951, Vol. XV:67–77.

———. "The Study of Race." *American Anthropologist* 65:521–532, 1963.

Watzman, Herbert M. "Question of Human Origins Debated Anew as Scientists Put to Rest the Idea of a Common African Ancestor." *Chronicle of Higher Education*, September 16, 1992, pp. A7–A8.

Weidenreich, Franz. "Facts and Speculations Concerning the Origin of *Homo sapiens*." *American Anthropologist* n.s. vol. 49 (2):187–203, 1947.

———. "Intepretations of the Fossil Material." In W. W. Howells, ed., *Early Man in the Far East*. Philadelphia: Wistar Institutes Press, 1949, 1:149–157. Reprinted in W. W. Howells, ed., *Ideas in Human Evolution: Selected Essays 1949–1961*. New York: Atheneum, 1967, pp. 466–472.

———. "The 'Neanderthal Man' and the Ancestors of *Homo sapiens.*" *American Anthropologist* 45:39–48, 1943.

Wheeler, David L. "U. of Md. Conference That Critics Charge Might Foster Racism Loses NIH Support." *Chronicle of Higher Education*, September 2, 1992, p. A6.

Wilson, Allan, and Rebecca Cann, "The Recent African Genesis of Humans." *Scientific American* 266 (4):68–73, 1992.

Wilson, James Q., and Richard Herrnstein. *Crime and Human Nature*. New York: Simon & Schuster, 1985.

Wilson, L. G. *Sir Charles Lyell's Scientific Journals on the Species Question*. New Haven: Yale University Press, 1970.

Wolpoff, M. H., Alan G. Thorne, Jan Jelinek, and Zhang Yinyun. "The Case for Sinking *Homo erectus*: 100 years of Pithecanthropus Is Enough!" Paper delivered at conference "100 years of Pithecanthropus," Senckenberg, Frankfurt am Main. To be published in Jens Franzen, ed., *100 Years of Pithecanthropus—The* Homo erectus *Problem*. Frankfurt am Main: Courier Forschungs-Institut Senckenberg.

Zangwill, Israel. *The Melting Pot*. 1908.

Index